The Technical Challenges and Opportunities of a United Europe

The Technical
Challenges and
Opportunities of a
United Europe

Edited by

Michael S. Steinberg

Barnes & Noble Books
Savage, Maryland

First published in the United States of America in 1990 by
Barnes & Noble Books
8705 Bollman Place
Savage, Maryland 20763

Library of Congress Cataloging in Publication Data

The Technological challenges and opportunities of a united Europe /
   edited by Michael Steinberg.
      p. cm.
   Includes bibliographical references.
   ISBN 0–389–20900–7
    1. Technology and state—Economic aspects—European Economic
Community countries—Congresses.   2. Technology transfer—Government
policy—European Economic Community countries—Congresses.
3. European Economic Community—Congresses   4. European Federation–
–Congresses.   I. Steinberg, Michael Stephen.
HC240.9.T4T4   1990   89-77848
338′.064′094—dc20   CIP

Printed in Great Britain

Contents

Part IV The Tripartite Relationship

Part V The EC and the Third World

List of Contributors

Michael S. Steinberg — Vice President of the Institute of European Studies, Chicago; Director of the Institute of Asian Studies, Chicago.

G. Edward Schuh — Dean, Hubert H. Humphrey Institute of Public Affairs, the University of Minnesota. Prior posts include: Director of Agriculture and Rural Development, The World Bank, and Professor and Head, Department of Agricultural and Applied Economics, University of Minnesota.

Donald J. Puchala — the Director of the Institute of International Studies at the University of South Carolina.

Klaus Kammerer — Instructor at the Verwaltungs-und Wirtschaftsakademie in Freiburg in Breisgau, an Academic superior counselor in Mathematics in the Economics Department at the University of Freiburg and an Instructor in the IES/EEC program in Freiburg.

Seamus O'Cleireacain — Associate Professor of Economics at the State University of New York, Purchase, Associate Director of the Center for Labor-Management Policy Studies at the City University of New York Graduate School and Director of the US—EC Seminars at the Institute on Western Europe at Columbia University.

Pierre-Henri Laurent — Professor and Chair of the History Department at Tufts University, and sometimes Adjunct Professor of Diplomatic History at the Fletcher School of Law and Diplomacy at Tufts University.

Alain Dumont — Professor of Business Strategy and International Business and Director of the International Program at Centre HEC—ISA near Paris.

Christopher Coker — Lecturer on International relations at the London School of Economics and Political Science.

Gavin Boyd — Professor of Political Science at Rutgers University, Newark New Jersey

Gregory O'Connor — Director of Regulatory Affairs at the United States Department of Commerce.

Lawrence G. Franko — Professor of Finance and Business Strategy at the University of Massachusetts, Boston.

C. Ford Runge — Director of the Center for International Food and Argricultural Policy and Associate Professor Department of Agricultural and Applied Economics, the University of Minnesota, where he also holds an appointment in the Hubert H. Humphrey Institute of Public Affairs.

Susumu Yamakage — Associate Professor of International Relations at the University of Tokyo.

Carol Cosgrove — Director of CTA Economic and Export Analysis Ltd. and Research Fellow at the University of Reading.

Carol Armistead Grigsby — Deputy Director of the office of Multilateral Financial Institutions in the US agency for government policy toward the World Bank and related development financing institutions.

Preface

The material in this book was first presented at a workshop organized by the Institute of European Studies and the Western European Area Studies Center of the University of Minnesota from 14–16, April 1989. The workshop was organized as a follow-up to three faculty development seminars held by the Institute in June 1982, 1983 and 1984.

The early 1980s was a period of apparent stagnation in the European Community and a time when American scholarly interest in the EC had flagged. One or two of the speakers at the Institute's seminars expressed uncertainty about the future of the market. The general consensus did not anticipate a breakthrough in the foreseeable future. We are now aware that in the early 1980s seeds of the 1992 movement were planted at meetings of the European Council, begining with the Copenhagen meeting of December 1982 when the representatives agreed to adopt measures that would 'reinforce the internal market'.[1]

The movement reached its concrete stage in June 1985 with the Commission's publication of the White Paper, 'Completing the Internal Market', which was drafted by EC Commissioner Arthur Cockfeld. By the end of 1985, the EC governments had signed the Single European Act which, when it went into force in 1987, substituted weighted voting for the Council's unanimity requirement in most areas relating to the internal market. The Single European Act also committed the signatories to fulfilling the objectives of the White Paper by 1 January 1993.

The 1992 movement has invigorated the Community and arrested the attention of the rest of the world. The United States, until recently preoccupied with Japanese competition and the rise of the newly industrializing entities (NICS), has been forcefully reminded that the Pacific basin is not the only realm of potential growth. As 1992 draws closer, the EC is gaining accelerated attention from the United States government, the business community and the academic world.

The Minnesota workshop brought together representatives of these three sectors. The speakers hailed from the United Kingdom, France, West Germany, the United States, Canada and Japan. The Contributors to this book represent a number of different viewpoints. They share the general view that the 1992 movement is a process, the ultimate outcome of which is unpredictable.

The new unity in Europe represents an opportunity for accelerated economic growth for Europe and for much of the rest of the industrial world. It remains uncertain how effectively this moment will be seized.

The chapters focus on a number of important issues. The first section is devoted to the keynote address delivered by G. Edward Schuh at the EC Workshop which brings together a number of the strands presented in this collection. He views the EC's evolution as a prime example of a world-wide trend to regional economic blocs, attributing this development to the vast expansion of the international capital market. He further associates with it a tendency toward devolution of economic management within nations to smaller political units similar to American states. Finally he reviews a number of key issues relating to 1992 in such areas as trade, agriculture, and monetary and fiscal policies.

In the second section, the authors look at the impact of the 1992 movement on Europe. Donald Puchala sets the stage of this section with an overview of the economic and political meaning of 1992. Klaus Kammerer explores the impact of international deregulation upon the German transport industry, depicting the promise and the dislocation that is likely in this important sector and, implicitly, in others. Seamus O'Cleireacain, an Irish-American scholar, identifies the dislocation problems that labour will face after 1992 and the initial steps taken in the direction of a trans-national labour market. The long-range impact of 1992 is especially unpredictable and problematical in the social area.

The technological challenge by the United States and Japan, discussed in the third section, was one major factor that led to the 1992 movement. In his chapter, Pierre-Henri Laurent chronicles the history of the European efforts to forge a common technological effort. Alain Dumont discusses the current state of technological cooperation in Europe, both as a result of EC and government initiatives and efforts that have come increasingly from the private sector in the last five years. The 'merger mania', in preparation for 1992, promises greater chances for technological progress. Christopher Coker is less optimistic about the prospects for Europe to cross new frontiers in military technology. American predominance in this area is protected by its far greater expenditures on research and development and the United States' increasing reluctance to share its military technology. The United Kingdom has failed to exploit its special relationship with the United States and increasingly plays a subsidiary role in defence technological development. The EC, excluded by the Treaty of Rome from national security matters, does not provide a framework for European cooperation in this area.

The fourth section of this collection discusses the tripartite relationship between the EC, the United States and Japan. Gavin Boyd offers a summary of the EC's interdependencies with the United States and Japan and looks at the United States–Japan relationship from the vantage of the EC. He assesses the difficulties that Europe continues to face in developing a common policy in its relations with the other economic superpowers. Gregory O'Connor then describes the United States' government efforts to help American business prepare for 1992. Lawrence Franko paints a pessimistic picture from the United States' perspective of European and American multinationals' efforts to take advantage of European unity. He suggests that transnational mergers within Europe will make European multinationals increasingly competitive in the American market. At the same time, in many sectors, American multinationals have been on the retreat within Europe. Ford Runge discusses the issue of trade policy reform in the sphere of agriculture. The elimination of agricultural barriers in Europe could lead either to a new protectionist bloc or to new opportunities for reform through the GATT. Susumu Yamakage expresses the Japanese concern that the emerging Europe will be more protectionist than it is today. While Japanese interests will be best served by an open, free-trading world, in self-defence Japan may be forced into a regional economic bloc in East Asia, in the Pacific basin or with the United States.

The final section is devoted to the EC's relationship with its partners in the third world, the ACP (African, Caribbean and Pacific states). Carol Cosgrove demonstrates that the Lomé Convention has protected markets of the ACP countries in Europe for traditional products, but has not contributed to diversification and expansion of ACP economies. The ACP countries fear that 1992 may worsen their situation and are pushing for further safeguards and developmental resources. Carol Grigsby discusses the growing importance of the EC in development aid to the ACP countries in Africa and suggests the possibility that the EC will adopt an independent, constructive role distinct from that of the World Bank.

Beginning with its origins in the Organization for European Economic Cooperation in the post-war period, and the establishment of the EEC with the Treaty of Rome on 1 January 1958, the movement toward European Economic unity has looked both inward and outward: Inward to the creation of a commercial and industrial entity that can provide a base for economic growth and expansion that can effectively compete with the continental powers like the United States, the Soviet Union and more recently Japan; and outward as a model for a world in which national boundaries become less important as a barrier for trade. The 1992 movement represents a new stage in this development. It could lead, as Susumu Yamakage warns, to a world of competing protectionist blocs. It can also contribute to the further opening up of world commerce. The efforts in the 1992 movement to eliminate non-tariff barriers, especially in the principle of mutual recognition of standards, point the way to the lowering of these barriers on a world scale. Whatever the course that the EC adopts, the consequences are likely to be felt well beyond the borders of the Community.

This book and the workshop it was based upon are themselves, the product of an international effort. Funds for the workshop were provided by the German Marshall Fund, the EC Delegation to the United States, the University of

Minnesota and the Institute of European Studies. William Gaines, President of the Institute of European Studies, has had a long-term and abiding interest in the EC that has inspired many efforts in this area including the IES undergraduate programme first run in 1982 in Freiburg, the three faculty seminars and this workshop. Contributors were assembled through the combined efforts of Bernhard Büchner of the University of Freiburg and IES–Freiburg, Alfred Fontenilles of HEC and of IES–Paris, Robin Simon of IES–London, Pierre-Henri Laurent of Tufts University and Kim Munholland of the University of Minnesota. Professor Laurent is responsible for much of the conceptual framework for this book. Professor Munholland provided the university setting that successfully brought the speakers together with a stimulating audience of faculty from IES affiliates and associates. Linda Brienzo of the University of Minnesota was responsible for local arrangements. Karen Hartwig of IES ably assisted with both the conference arrangements and with the assembling of this book. The Board of Governors, Institute of European Studies (and the new Institute of Asian Studies) deserves thanks for nurturing and supporting the IES–IAS faculty development programme and the scholarly interests of its staff.

Michael S. Steinberg
Chicago, August 1989

Note

1. Michael Calingaert, *The 1992 Challenge from Europe: Development of the European Community's Internal Market*, Washington, D.C., National Planning Association, 1988, p. 8.

Part I Introduction

1 The European Community: looking towards 1992

G. Edward Schuh

EC–92 will be a very important event if it does in fact come off, whether we think of it economically, politically or historically. And there is every reason to believe it will indeed come off. The momentum is great. An economically integrated European Community will be an economic powerhouse. It will also have very important political implications on the international scene. Both the economic and political implications can and probably will be very far reaching for other parts of the world as well.

There are five issues I would like to discuss: these are critical issues either in bringing about the establishment of the new Community or in sustaining its continuing success.

The bifurcation of economic policymaking and implementation

There is an important process of change taking place in the international economy both in *where* economic policy is being made and in terms of the level at which it is being implemented. These changes are coming about because of the increased openness of national economies. This increased openness is the consequence of the growth in international trade relative to global GNP, a process that has been going on throughout the post-World War II period. It is also a consequence of the emergence of a huge international capital market, a process which started in the mid-1960s and extended through the 1970s. This international capital market ties national economies together in ways that are every bit as important as those through international trade. More importantly, it links macroeconomic policies together in ways they haven't been linked since the days of the gold standard.

The need for change comes about because economies that are increasingly open are also increasingly beyond the reach of national economic policies. Because of this, there is a tendency for economic policymaking and policy implementation to shift in two opposing directions. One part of it shifts to the international level in the form of the codes, disciplines and rules and organizations such as the General Agreement on Tariffs and Trade (the GATT); the other part shifts to the state and local level.

In both shifts, the character of the policy changes as it changes location. For example, with closed national economies, national governments have a great deal of control over their macroeconomic policies. They also tend to have sectoral policies which focus on individual sectors of their economy and which operate primarily through interventions in their product markets. Perhaps the best example of such sectoral policies are the agricultural commodity programmes which prevail in both the United States and the European Community.

As economic policymaking shifts upward to the international level, national governments lose control of these sectoral policies. This is made evident in the case of the United States by the large Treasury costs the agricultural commodity programme have incurred in recent years, with the agricultural sector remaining in the doldrums despite large expenditures. Both the United States and the European Community protect their agricultural sectors by means of barriers to trade. But these eventually become an international issue and subject to the regulations imbedded in the GATT. In this sense, the 'regulation' of the commodity markets shifts upward from the national level to the international level.

At the same time that this occurs, a great deal of the remainder of domestic sectoral policy shifts downward to the state and local level. The character of this policy also changes. For example, it is not feasible for state governments to support domestic prices above market clearing levels or to acquire stocks in order to support prices for farmers in their states. Instead, income policies become focused more directly on the factor or input markets and other aspects of resource policy come to the fore. Ultimately this is an advantage in an efficiency sense since policy can more properly reflect the heterogeneity in resource endowment. There is loss in equity as this shift takes place, however, at least when viewed from the national level, since within a range per-capita incomes will reflect underlying resource endowments.

This decentralization of economic policy is already rather far advanced in the United States. State governments have grown rapidly over the last decade, with higher salaries being paid and an improvement in the quality of services rendered. Unfortunately, in the case of the United States, this process is all too often referred to as the 'Reagan Revolution'. In point of fact, the process is a reflection of far more basic economic forces. That it is a reflection of more basic economic forces is evidenced by the fact that the same process is occurring in other parts of the world. Brazil's recently promulgated constitution, for example, institutionalizes the decentralization of economic policy to the state and local level. At the other extreme, the various examples of economic integration we are witnessing on the international scene, such as the Canada–United States Free Trade Agreement, are cases in which policy is shifting up to the transnational or international level.

EC–92 needs to be understood in this same context. It is a case in which policymaking and implementation is shifting from the national level to the international level. Nationally established barriers to trade and to the mobility of capital and labour will be eliminated and will shift upward to the level of the European Community.

Interestingly enough, when this occurs, current problems at the 'state' and local level will be exacerbated. The 'states' in this case will be what are now the nations. As we will see below, adjustment problems will be severe as the integration process evolves. But there will also be a need for a major transformation of policy at the level of what is now the national governments. This will pose major challenges of institutional design and reform.

Economic integration as a means of trade liberalization

EC–92 constitutes an important example of economic integration. There are a number of such initiatives on the international scene, with the recently concluded free trade agreement between the United States and Canada being one of the most significant. The volume of trade between these two countries is the largest of that between any two countries in the world.

In my view, such economic integration is one of the most powerful forces for trade liberalization on the international scene and has the potential for being the driving force for a great deal more liberalization—far more than can be obtained by means of the multilateral trade negotiations currently taking place in Geneva.

Consider the Canada–United States Free Trade Agreement as a case in point. Given the potential volume of trade that may be involved, it already constitutes a significant act of trade liberalization. But it could also be a considerable force for further liberalization. All that would be required is that the United States and Canada say to all other countries, 'anybody who wants to join can do so so long as they are willing to abide by the same rules'. Given the attractiveness of this huge market, many countries such as Brazil and Mexico might be willing to join, especially if provision were made to spread the process of integration over a period of years.

Other, more limited, suggestions for the extension of the Canada/United States agreement have already been made. Some have suggested, for example, that Mexico be brought into the agreement, with the result being a North American common market. This is not as far-fetched an idea as it might seem on the surface. Mexico has already taken large steps to reduce its barriers to international trade. The greatest barrier to further integration with the United States and Canada is Mexico's continuing barriers to foreign capital. This is probably not as difficult to remove now as it would have been in the past.

Another suggestion has been that Japan join the Canada/United States agreement. Japan's opening of its capital markets probably makes such an integration more feasible than it was, say, ten years ago. However, there would still need to be considerable trade liberalization on the part of the Japanese, and political difficulties in the United States could also be considerable. In any case, these examples of potential integration, with an associated liberalization of trade, illustrate the potential of economic integration to promote freer trade.

EC–92 itself is an important example of how economic integration can be a trade-liberalizing force. The reduction of internal barriers to trade will by itself constitute a substantial amount of trade liberalization. The reduction of barriers to the mobility of capital and labour will be of historic significance. When was the last time such a significant opening of factor markets occurred?

There is a risk in the case of EC–92 which concerns many outside observers. That risk is that the new Community will raise its external barriers to trade at the same time it reduces them internally. If it should do this, it would likely lead to an international trade war of unprecedented proportions. The result could well be a significant increase in barriers to trade globally rather than the net reduction which is expected from the proposed integration.

It is not obvious at this time what direction the Community will take. Its persistent resistance to any reduction in the protection of its agricultural sector does not bode well for a reduction in, or even the sustaining of, present external barriers to trade. On the other hand, the pressures for undertaking EC–92 in the first place came from firms in the private sector who recognized that larger domestic markets were needed if they were to realize economies of scale and thus be efficient competitors on the international scene. Political leaders also recognized the need for integration for these same reasons. Thus there are grounds for hope that the establishment of EC–92 will on net lead to a reduction in barriers to trade.

Political leaders in other parts of the world, however, need to be alert to the possibility of back-sliding on the part of the EC. Academics also have a responsibility not only to be alert to this possibility, but also to call it to the public's attention and to help design institutional arrangements which make it possible to move forward with trade liberalization, rather than to slide back into more protectionism. If present levels of protection can be sustained, and possibly reduced, the new EC could eventually do as was suggested above for Canada and the United States—extend an invitation to other countries to join it so long as they abide by the same rules.

The EC–92 and the developing countries

The realization of EC–92 could be a significant positive event for the developing countries. If successful, it will constitute a large market for these countries, with the expected rise in per capita incomes from integration giving further impetus to market expansion. Given that the EC will emerge as the second largest economic power on the international scene, it also has the potential to be an important source of capital and foreign aid for the development of these countries.

Whether this potential is realized will depend very much on the conditions under which the EC–92 is finally established. For example, perhaps the most important issue is what the Community does about its agricultural policy. Its Common Agricultural Policy (the CAP) already has done great damage to the developing countries as protectionist barriers have progressively closed out potential markets for these countries. Sugar is an important example, as is beef. These are commodities for which the countries of the Community at one time offered significant markets for producers in the developing countries, but which over time have been closed out by means of protectionist barriers. Other examples could be cited.

Perhaps even more damaging has been the export subsidy war the EC and the United States have been engaged in over their agricultural trade. Protectionist measures in both have caused domestic supplies to outpace domestic consumption and what could be disposed of by means of regular commercial exports. The result has been the accumulation of large, costly stocks and the need to dispose of these by some means. The competitive use of export subsidies to dispose of the stocks abroad has caused the international prices of commodities so affected to be lower than they would otherwise have been.

For developing countries that are exporters of these commodities, this depression of prices has been a serious problem, especially for countries such as Argentina which have a comparative advantage in agriculture and which need to earn foreign exchange to service their considerable foreign debt. For countries that are net importers, on the other hand, the export subsidies constitute income transfers. But those income transfers may not be a full blessing. Clearly their consumers benefit. But their producers may well suffer significant income losses, and already severe domestic adjustment problems may well be exacerbated. Unskilled labour is forced prematurely out of agriculture and piles up in urban centres where it is unemployable. The result can be a significant loss in production potential for developing countries.

Even the reduction in price for consumers may not be without its costs. Depending on the import elasticity of demand for these commodities, the import bill may well burgeon at the very time the country needs to earn a large trade surplus to service its foreign debt and to pay for the imports it needs to further its economic development. Given that the imports of the subsidized agricultural commodities tend to have close domestic substitutes, the likelihood is that the import demand will be elastic. Thus, the reality is that these export subsidies exacerbate the problems of the developing countries rather than being of net benefit to them.

Should the Community rationalize its agricultural policies at the same time it moves towards fuller economic integration, it would be of considerable benefit to the developing countries. Should it fail to do so, the damage to the developing countries would continue into the future and possibly increase over time.

The same problem applies to other goods and services. Many developing countries are developing the potential to be international competitors in labour-intensive manufactured goods. With low external barriers to the trade in these goods, the Community could constitute an important market for these countries. But if it raises its external barriers in the process of furthering its own economic integration, this potential would be sacrificed.

On the issue of foreign aid, the Community has done better than the United States, especially if performance is measured by the share of its net national income it provides as foreign aid. But this is consistent with the late Harry Johnson's dictum that foreign aid is the soft option when it comes to the aid versus trade trade-off. Foreign aid is provided in lieu of freer trade. Everybody would be better off, of course, if the emphasis were put on freer trade instead of foreign aid.

In conclusion, the creation of EC–92 has the potential to be of considerable positive value to the developing countries. The economic growth which it promises to generate can be the source of expanding markets for these countries, and that same economic growth may well provide the means for larger amounts of foreign aid. The key issue will again be how much external protection the new Community has once it is established. If protectionist barriers are high, and foreign aid is provided in lieu of trade rather than in addition, that potential will be sacrificed.

Monetary issues

Monetary issues have already come to the fore as a difficult political issue in the establishment of EC–92. One issue focuses on the exchange-rate system and in particular whether Great Britain will become a part of the European Monetary System. The second issue involves the decision of whether or not to establish a European Central Bank. These issues are not unrelated. How they are resolved may well be the central issue around which the success or failure of the more well-integrated Community is determined. It will also determine whether political integration becomes part and parcel of economic integration.

The choice of exchange-rate regime within the Community will be the key issue. It currently is the most controversial issue before policymakers at this time, as well it might be. Most Europeans seem to have an abhorrence of flexible exchange rates, preferring instead the fixed exchange rates of the old

Bretton Woods system. As a result, shortly after the rest of the developed countries shifted to flexible exchange rates in the 1970s, some of the European countries established the European Monetary System. This system essentially established a quasi-fixed exchange-rate system, with currency values among member nations permitted to fluctuate within a narrow band, often referred to as the 'snake'.

This system, like the old Bretton Woods fixed exchange-rate system, has the advantage of forcing discipline in macroeconomic policy on the part of participating countries. Exchange rates are kept within the band by changes in domestic monetary and fiscal policies. As long as one or more countries pursue conservative monetary and fiscal policies, the other countries are forced to do the same if they are to remain a part of the system.

This system is not an unmixed blessing, however. In the first place, as the above description implies, national governments lose a great deal of sovereignty over their domestic macroeconomic policies. That is why Margaret Thatcher resists joining the system. In addition, the exchange rates which prevail in such a system are seldom ideal for everybody. In fact, the likelihood is that the prevailing exchange rates are less than optimum for everybody. That is because the rates that prevail are essentially a reflection of average monetary and fiscal conditions for the group and nobody may be at the average.

Experience with the European Monetary System is consistent with that proposition. Western Germany, which tends to pursue conservative monetary and fiscal policies, has chronically tended to have an undervalued currency within the system. Consequently, it has persistently tended to run a surplus on its trade accounts. At the other extreme, Italy tends to be considerably less than conservative with its monetary and fiscal policies. As a result, it has tended to have an overvalued currency with all that that implies about its balance of trade and balance of payments.

If the Community should persist in its attempt to have a fixed exchange-rate regime internally, the establishment of a European central bank will become almost an imperative. The establishment of such a central bank, in its turn, will almost inevitably require some degree of political integration or union.

There is an alternative to such a fixed exchange-rate regime, however. It is to have each country pursue a flexible exchange-rate policy. In this way, each country would be free to pursue its own monetary and fiscal policies, and to preserve more of its national sovereignty in economic policymaking. It is because the United States and Canada both have flexible exchange-rate regimes that the issue of a need for a common central bank never came up as the free trade agreement was negotiated or that there has been no discussion about possible political unification of the two countries.

For successful economic integration to continue under a flexible exchange-rate regime, it would be necessary for member countries to pursue macroeconomic policies that were reasonably consistent with each other. If they should not do so, the disparities in policy will induce large capital flows among the countries which in turn will induce large swings in real exchange rates. These in turn will generate pressures for protectionism to offset the distortions in exchange rates and the eventual breakdown of the integration.

It may well be that some political leaders in Europe sense this danger and thus press for a system along the lines of the present European Monetary System. Given that there is already a significant desire for political unification on the part of some leaders, insistence on a fixed exchange-rate system is one means of furthering that cause.

In concluding this section, it is important to emphasize the extent to which choices about economic arrangements and institutions have very important political implications. How these choices are worked out over the next two years will be one of the most fascinating aspects of the effort to establish a more well-integrated Community.

Adjustment policy

The establishment of effective adjustment policies will be the key to the eventual success of EC–92. Even if there should be no monetary union, and national governments instead pursue flexible exchange-rate policies, the adjustments from trade liberalization itself can be quite large and significant. If there should be monetary union around some kind of a fixed exchange-rate system, effective adjustment policies will be an imperative. The problem is that the disparities in economic development among the various countries and regions of the Community are quite great.

The opening of both the labour and capital markets at the same time that trade is liberalized will help reduce the need for and facilitate needed adjustments. Labour will be free to migrate to new employment throughout the Community, and capital will be free to seek out areas in which wages are low. However, there will still be a significant need to facilitate the adjustment of labour. Much of the labour from disadvantaged regions will not have the skills required for expanding employment in other regions, and even if capital should flow into the regions that are disadvantaged, there will still be a need for skills for the new economic activities that will be created.

Successful adjustment policies start with a strong educational system, especially at the primary and secondary levels. Cognitive skills are the key to having workers who can learn about and adjust to new employment opportunities. Next on the list are training and retraining programmes that provide the skills that are needed in those economic activities that are expanding. Policies which help workers migrate by underwriting part of their moving costs and helping to sustain them in their new location until they are productively employed can also be quite helpful. Finally, an effective labour-market information system which helps inform members of the labour force where new opportunities exist, and which informs private firms where pools of unemployed workers are located, is fundamental to an effective adjustment policy.

An important issue which arises in this context is whether it is more desirable for labour to do the migrating or for capital to go where the labour is located. There is no easy answer to this question and a lot of factors need to be considered. However, I have argued elsewhere that outmigration from a region imposes significant negative externalities on the supplying region, largely associated with the loss of human capital. It turns out that migration tends to be highly selective, with the migrants tending to be younger, more well educated and more enterpreneurial than those remaining behind. The loss of this human capital is hardly an effective way of bringing about equilibrium in the disadvantaged region. Thus I argue that governments can contribute to a more efficient allocation of national resources, and bring about adjustment to changing economic conditions more rapidly by subsidizing the relocation of economic activities to disadvantaged regions rather than by facilitating outmigration from such regions.

Concluding comments

I would like to conclude by making two points. The first is to repeat what I said at the beginning. The successful realization of EC–92 is an event of enormous economic, political and historic significance. The implications for the Community itself will be quite great, but the implications are also of global significance, especially if political unification should eventually follow.

The second point is that the United States, both its political leaders and its citizens, needs to wake up to the significance of EC–92. The realization of a new European Economic Community has very great implications, both economic and political, for this nation.

Part II The impact of the 1992 movement on Europe

2 The economic and political meaning of Europe 1992

Donald J. Puchala

When, in 1985, the member states of the European Community (EC) were attempting the rewrite and strengthen the Treaty of Rome, the Common Market's constitution, Prime Minister Margaret Thatcher thought rather little of the exercise and accordingly described the whole enterprise as 'airy fairy'. Later that same year, when in fact the Community had substantially amended the Treaty, the British, while cooperative, remained sceptical. A British diplomat remarked at the time that the whole reform of the EC looked to him to be 'about as solid as Gruyère cheese'.[1]

Gruyère is in fact a rather hard cheese, but this is not what the Englishman meant in making his analogy. His expectations about 1992, and indeed the expectations of many people within and without Western Europe, were very modest in 1985. Needless to say, these modest expectations were also quite reasonable in light of the history of the European Common Market and in light of many people's images of Western Europe in the 1980s. In the first years of the current decade European countries—reeling from the double impact of the second oil price shocks and the global recession provoked by early Reaganomics—were still experiencing negative growth in GNP and soaring levels of unemployment. The entire Western world was in economic trouble, and Europeans were sunken in 'Europessimism'. Many were convinced that Western Europe had fallen impossibly far behind in technological development and many had already accepted that the high-tech future probably belonged to Japan. Economic conditions had improved by 1985, but, generally speaking, the relaunching of the drive toward greater European economic unity did not come at a very promising time. For this reason, the new beginning did not attract very much attention either within Western Europe or without. Yet the inauspicious start of the 1992 movement is one of the factors that makes its dramatic progress to date all the more impressive.

There is now every reason to believe that much of what the members of the EC want to accomplish under the rubric 'Europe 1992' will in fact be accomplished, if not by the end of 1992, then probably progressively in ensuing years. More interestingly, and perhaps ultimately more importantly, the momentum that the enthusiasm over 1992 has unleashed is leading Western Europe beyond the stipulations and expectations of the Single European Act and toward a greater unity than even the 1992 enthusiasts had imagined. In a series of BBC lectures delivered many years ago, the venerable British scholar Sir Andrew Shonfield described the course of post-war intra-European international relations as a 'journey to an unknown destination'.[2] The destination remains unknown, or at least uncertain, but the journey has started again and it is transforming the old continent.

1992 and the completion of the Common Market

Let us recall that the European Common Market scheduled to be completed by 1992 was started in 1952 with the merger of the coal and steel industries of six countries—France, West Germany, Italy, Belgium, the Netherlands and Luxembourg. In 1958 these same six countries integrated their economies into a customs union, wherein tariffs and quantitative restrictions on trade were eliminated, a common external tariff was erected, agricultural and fisheries policies were supranationalized under the authority of a European Commission that operated from Brussels and early steps were taken to harmonize tax systems. In 1979 several of the Community countries joined an European Monetary System (EMS) where they pledged to keep their currency exchange rates closely linked, and even before that some national, international and business accounting began to use a standardized European unit of account—the ECU. Meanwhile the size of the EC has been enlarged from six, to nine, to ten and finally to twelve. The initiatives concerning 1992 therefore are best understood as a continuation of this European integration process. They do not come 'out of the blue'.

The White Paper drafted by EC Commissioner Arthur Cockfield in 1985 and the Single European Act that adopted most of it as a blueprint for reforming the EC, set three principal objectives for the completion of the common market.[3] The plan called first for removing physical barriers at Community borders—that is customs houses, transhipment points and all of the delay-causing and money-costing routine monitoring and paper work that interrupts the smooth flow of intra-Community commerce. Because every member country has its own requirements for customs declaration, bills of lading and other forms of certification for cargoes moving across its frontiers, typical haulings through the EC consume needless hours of stopping and waiting time and reams of redundant paper, all at the expense of speed and efficiency. The programme for 1992 called for streamlining border-crossing procedures.

Second, the plan for 1992 sought the removal of countless technical barriers to the flow of goods and services across EC national borders. Here the focus was on almost everything that made the Common Market uncommon: differing national product standards, labelling standards, health and veterinary restrictions, certification requirements, public procurement practices and countless other nationally idiosyncratic rules, regulations and devices that served deliberately or *de facto* as non-tariff barriers to Community economic intercourse. All of these were to be 'harmonized', 'approximated', 'mutually recognized', 'coordinated' or 'standardized' to render goods and services from anywhere in the EC acceptable in all member countries.

Third, differing national fiscal systems were to be nudged toward harmonization. Removing fiscal frontiers, or in other words neutralizing the trade-distorting effects of differing national tax structures, has long been an aspiration of the EC. Important steps were taken in the 1960s and 1970s with the establishment of a common sales tax, the VAT or tax on value added.[4] By 1992 the Community plans to standardize the rates of this tax transnationally as well as to approximate a number of national excise taxes. Some in the EC are also looking beyond 1992 to the harmonization of income taxes and toward the transnational coordination of fiscal policies.

The three-pronged assault on physical barriers, technical barriers and fiscal barriers to the free flow of international commerce in the EC is currently in full swing. Procedures for harmonization, approximation and standardization are being written into a bevy of so-called 'directives', or acts of EC legislation, which, after approval by the European Council of Ministers, become mandatory for the member governments. Community authorities in Brussels are currently at work on some 275 such directives, all of which are expected to be in force by 1992. So far, about 140 of these 1992 directives have found their way completely through the complicated EC legislative process.[5]

Some results of completing the Common Market

There are two questions that need to be raised and answered about the future of Europe 1992. The first is, what will happen as a result of the adoption of the 275 directives? The second is, what will happen as a result of the overall 1992 initiative?

Most of the analysts who are currently monitoring the EC legislative process concur that the majority of the 1992 directives will ultimately be adopted, and that these will be enforced by the different member states. No one is insisting that everything will be in place by 31 December 1992, but a great deal will. The overall result of this will be to facilitate greatly transnational commerce in the EC—in effect, to complete the Common Market as planned.

With the directives implemented it will be possible:

— to cross national borders for commercial purposes with minimal customs house delays, minimum inspection and reporting procedures and greatly reduced recording requirements;

— to sell everywhere in the Common Market products produced anywhere in the EC, without requirements for separate and different marketing authorizations and certifications;
— to provide everywhere in the Common Market services originating or certified anywhere in the EC, including contractual services performed for national governments and other public authorities;
— to engage in professions everywhere in the Common Market if licensed or certified to engage in these professions in any member country;
— to expect similar structures and rates of indirect taxation everywhere in the Common Market.

As a result of the international economic integration that the directives will foster in the years after 1992, the European Community can expect a sharp increase in intra-community trade (in both products and services) and ensuing positive impacts on national incomes. The European Commission has estimated an immediate boost in trade by over 20 per cent and a quickly registered 6–7 per cent rise in total Community GNP.[6] These increases in intra-regional economic activity will be differentially felt in different countries and in different industries, but expert consensus now predicts that no countries will be adversely affected. Nationally speaking, everybody is expected to gain something in the aggregate from the initial economic effects of 1992.

Preparing for 1992 and then operating most effectively on the new Common Market that 1992 will create, are going to require considerable adjustments in business strategies and behaviour. Advantage will accrue for example to firms that are able to realize economies of scale. Advantage will accrue to firms that are linked into distribution networks in other countries. Advantage will accrue to producing products that are unaffected by culturally conditioned consumption habits or national tastes. In many ways, advantage will accrue to large, multinationally owned, organized and operating companies. For this reason there is now a veritable stampede toward bigness in Western Europe.[7] Offensively intended cross-national mergers are taking place, as well as defensively intended national mergers. The merger mania is likely to become more feverish over the next few years.

There is also the beginning now of a predictable rationalization of production and distribution also designed to take advantage of the emerging continental market. Manufacturers are seeking to locate their production facilities in the lowest-cost regions of the EC, with the confidence that barriers to cross-border transport will be minimized and that production standards will be harmonized. At the moment, there is considerable movement toward locating plants in Spain and Ireland in order to take advantage of lower wage rates in these countries. There is also a not unreasonable expectation that the efficiencies realized through economies of scale, rationalized production and distribution, enhanced investment in research and development stemming from bigness and keener marketing strategies required by keener competition all will result in broader choice and lower prices for Western European consumers. These new commercial conditions could also result in enhanced Western European competitiveness *vis-à-vis* third countries.

As might be expected from such a dramatic set of changes as 1992 will bring, there will be economic dislocation, and there are going to be some losers. For one thing, the rationalization of production and distribution is going to create short-term imbalances in the supply and demand of labour. There is a good possibility that some of the poorer EC countries—like Spain, Portugal and Ireland—will benefit from this; some of the higher-wage northern European countries may experience increased unemployment. There is not likely to be very much transnational labour migration because cultural and linguistic differences and differences among national social welfare regimes have rendered labour relatively immobile in Western Europe in the post-war era. Most of the labour movement in response to 1992 will take the form of rural to urban migration or agriculture to industry movement in the southern European Common Market countries. There is, however, likely to be a good deal of white-collar mobility as firms merge and move their production facilities to meet new market conditions.

There is some question now as to how small- and medium-sized firms will fare on the completed Common Market. One possibility is that they will benefit generally from the increased consumer demand that the completed Common Market will generate. They might also find their sales and revenues enhanced to the extent that they are suppliers to the multinationals. On the other hand, small- and medium-sized firms could be hurt as a result of the creation, expansion, scale efficiencies and pricing advantages of the giants. This appears particularly to be the case in retailing. A best estimate right now is that the distribution of costs and gains among differently sized firms will be sector or industry specific, and a good deal will depend upon firms' preparations for 1992 and their strategies for operating on the transnational market. Right now, it is information that is in greatest demand as businessmen seek to determine exactly how the 1992 directives will affect the market environment in which they are operating.

Beyond the Common Market and 1992

As impressive as the economic results of 1992 are likely to be, these predictable economic consequences are only part of what is likely to happen to the EC in the years ahead. There is a good chance that economic unification of an even farther-reaching nature than that planned for 1992 will emerge in the 1990s. Furthermore, many in Western Europe today believe that the 1990s will also see further steps toward political unity among the twelve.

Those who seek to push Europe toward greater unity beyond 1992 imagine a sequence of steps that if taken and successful will transform the EC into a loose confederation of states. It will not be a United States with the sovereignty of the separate states superseded and a federal government installed. Almost no one is imagining this any longer, and few actually believe that such a federation is desirable. What is desirable, and attainable, is a system of continuous, very close coordination among the national governments, facilitated and to some extent orchestrated by a set of central institutions. The function of such institutions would not be to legislate policy, but rather to frame and abet the consultative processes that promote agreements among the national governments. What is probably most feasible is an essentially inter-governmental arrangement, not unlike the one proposed by French President Charles de Gaulle in the early 1960s, that is *l'Europe des Patries*.[8]

Those who see the scenario of loose confederation unfolding, imagine first the completion of economic union among the twelve with the establishment of a monetary union, a common currency, a central bank and the introduction of an international consultative process that will generate Community-wide monetary policy.[9] This will be complemented by a Community-wide fiscal policy of sorts, even though taxing authority will stay largely in national hands. While the monetary union is being constructed, further steps will be taken to coordinate the foreign policies of the twelve on all possible major issues of world affairs. The seeking of a common foreign policy has been in progress since the early 1970s.[10] It is specifically mandated by the Single European Act and as such is part of the 1992 thrust. The Single European Act also brings the European Commission for the first time formally into the foreign policy process, which is called European Political Cooperative or EPC. This is significant because it adds the coordinating and consensus-building skills of the Commission to the EPC process and in so doing probably adds to its efficiency.

To date the EC have hammered out common policies on a broad range of international issues including Palestine and the Middle East, Afghanistan, Central America, Namibia, Cambodia, disarmament and chemical weapons, human rights, the protection of the global environment and relations with Eastern Europe. They enunciate these policies through appointed spokesmen, promote them through the United Nations and other multilateral channels and stand by them in separate bilateral dealings. The collective stands of the twelve project the Community as a power in world affairs. The world increasingly recognizes this. The EC governments and Community officials know it, and like it. Common EC positions frequently locate the twelve either between the United States and the Soviet Union on East–West or between the developing countries and the United States on North–South issues. These mediating stances nowadays are frequently points of convergence for international compromises, and therefore cast the EC in an important international broker's role. The twelve appear to have every intention of moving forward in their efforts to formulate Community-wide foreign policy and to project it out into the international system in the 1990s.

Questions of collective Western European defence centred in the EC are just now being seriously raised.[11] For those who imagine an eventual loose confederation of Western European states, the completion of a common defence policy is to be the crowning achievement, the last step. Presently, there is a good deal of thinking going on in European strategic analytical circles. It concerns the principles of a common defence policy in the 1990s and beyond, the meaning of Western European security in a possible post-cold war setting, the articulation of security policy and disarmament policy, and the relationship between NATO and EC collective defence. The dialogue is in a very preliminary phase, but it is in progress.

How likely is it that these further economic and political steps toward European unity beyond 1992 will be taken? The truth is that it is impossible to say. The last thirty years of efforts toward European unity have certainly not lived up to the expectations of those who wanted to build a political federation on the basis of their desires to perpetuate intra-European peace. Expectations have always been somewhat ahead of accomplishment regarding European unity and there is therefore some call for caution. On the other hand, the current relaunching has a great deal supporting it and pushing it forward that was missing earlier. For one thing, the British are on board. Despite Mrs Thatcher's opposition to bureaucratization, centralization and over-regulation at the Community level, she and many around her do nevertheless favour moving forward toward greater European unity. British support is important, partly because British opposition can be very destructive, and partly because the other major member states are already strongly in favour of pushing forward. Second, public opinion in most of the EC countries, continues to be very much in favour of moving farther, and faster toward greater political unity, albeit in a democratic manner.[12] In addition, the *détente* climate has opened opportunities for European global diplomacy which, if Western Europe is to make the most of them, make greater unity almost imperative. Finally, there is 'Europe 1992' itself: not the technicalities of it, but rather the psychology of it. Thus far 1992 has been a tremendous confidence-builder for the EC. As one United Nations diplomat put it, 'Europe is on a roll'. Just as surely as movement toward greater unity was impossible during the years of Euro-pessimism, not moving toward greater unity is almost impossible in the current climate of Euro-optimism. Atmospherics count in international relations, and right now European accomplishment is certainly in the air. In fact the European air is tingling over 1992.

Even Margaret Thatcher can no longer believe that 1992 is an 'airy fairy'.

Indeed, it is also a good deal more solid than Gruyère cheese. There is no question but that something is going to happen in the EC by 31 December, 1992, and even more is likely to happen in the years ensuing. Considerable adjustments are going to have to be made in the face of what 1992 is going to cause economically. These adjustments are in fact already being made by firms, by labour unions and other societal forces and by national governments. On balance the economic outcomes will be benign for the peoples of the EC (and probably also for Europe's external economic partners). But what is probably most significant about 1992 is that is has relaunched the movement toward Western European political unity. How far and how fast this movement will progress depends upon an array of factors, a number of which have to do with matters and modes of global relations that are largely beyond Western European control. But, right now it appears that the door is open to a greatly enhanced Western European role and influence in the world. It also appears that governments of the twelve have every intention of moving collectively on through this door.

References

1. Donald J. Puchala, 'Reforming the European Communities: 'Airy Fairy' or "Gruyère Cheese"', *Il Politico*, **LII**, 2, 1987, pp. 213–31.
2. Andrew Shonfield, *Europe: Journey to an Unknown Destination* (London: Penguin Books, 1973). For another earlier prognostication that is proving more durable as time moves on, see, Johan Galtung, *The European Community: A Superpower in the Making*, London, Allen & Unwin, 1973.
3. Commission of the European Communities, 'Completing the Internal Market', White Paper from the Commission to the European Council, COM (85) 310, Brussels 14 June 1985. See also the commentary on the White Paper in Michael Calingaert, *The 1992 Challenge: Development of the Community's Internal Market*, Washington, National Planning Association, 1988, pp. 20–9; 'Single European Act', *Bulletin EC*, Supplement, 2/86.
4. Donald J. Puchala, *Fiscal Harmonization in the European Communities: National Politics and International Cooperation*, London, Frances Pinter, 1984, ch. 2.
5. Commission of the European Communities, 'Third Report From the Commission to the Council and the European Parliament on the Implementation of the Commission's White Paper on Completing the Internal Market', COM (88) 134 Final, Brussels, 21 March 1988; Commission of the European Communities, 'A New Community Standards Policy', Brussels, 1988.
6. Calingaert, op. cit., pp. 66ff; Paolo Cecchini, *The European Challenge 1992*, Aldershot, Wildwood House, 1988, ch.1 and *passim*.
7. *International Herald Tribune*, 'Cross-Border Deals Point to Need for New Rules', 6 April 1989 p. 6.
8. Christopher Johnson, 'De Gaulle's Europe', *Journal of Common Market Studies*, , I, 2,

1963, pp. 154–72; John Pinder, *Europe Against de Gaulle*, New York, Praeger, 1963, pp. 31–46.

9. Elke Theil, 'From the Internal Market to an Economic and Monetary Union', *Aussenpolitik*, I/89, pp. 66–75.

10. Roy H. Ginsburg, *Foreign Policy Actions of the European Community*, Boulder, Lynne Rienner, 1989, pp. 55–116.

11. John S. Westerlund and Volker F. Fritze, 'The Franco-German Brigade, *Defense and Diplomacy*', **VII**, 6, 1989, pp. 24–8; Jacques S. Gansler and Charles Paul Henning, 'European Acquisition', *Defense and Diplomacy*, **VII**, 6, 1989, pp. 29–35; 62.

12. Commission of the European Communities, *Eurobarometer*, No.30, December 1988, pp. 45–51.

3 The integration of systems and non-systems—EC-92 and the West German transportation carriers

Klaus Kammerer

(translated by Michael S. Steinberg and Leo Van Cleve)

The meaning of trade policy

The *Wall Street Journal* once wrote that the success of integration hinged upon whether the British became the police, the French the cooks, the Italians the lovers, leaving everything to the Germans to organize; then heaven would be created in Europe. But, if it happened that the French became the police, the British the cooks, the Germans the lovers and everything was organized by the Italians, everything would then be so distorted that it would resemble hell. This American joke is perhaps very amusing for Americans. In Europe, however, the smile becomes increasingly more tortured as the 1992 deadline approaches. It is typical that shortly before the deadline after which some circumstances will change seriously, anxiety sets in. Brides are in the habit of crying on the morning of their weddings—at least Europeans notice this in Doris Day films.

It would be too complicated here to analyse completely which anxieties are justified, what results harmonization measures would have and how these harmonization measures will be carried through politically. This chapter will attempt, from a German point of view of course, to describe a part of a problem of a part of the market: namely the consequences that are beginning to appear for West German long-distance trucking and perhaps arriving at a few general criteria for the evaluation of the harmonization measures taken by the Commission, the Ministerial Council and the governments. Long-distance trucking is the first area in the service sector in which the Commission was forced to act and on which it is possible to draw conclusions for the overall development of organizational policy in the service sector.

The transport of goods plays a significant and complicated intermediate role in economic theory. If we divide the creation of value in an economy into two classes: (a) production of goods and (b) production of services, it is clear that the production of automobiles belongs to the production of goods and the work of a beautician belongs to the area of services. In which area should the transportation of goods be placed? Does it make sense to produce a good if it is not brought to a consumer? How can steel be produced if the raw material and the production element of coal are not delivered?[1] If the distribution of the goods is not carried out by the manufacturer concerned but by an independent transporter, then it is actually a service that must be taken into account. The intermediate position of freight transport between the production and the service sectors is in fact the reason that for thirty to forty years the Soviet Union, for ideological reasons, did not include services in the social product but did include freight transport in the state creation of value along with the production of goods.[2] This situation has also led to most governments having transport ministries, and to the fact that within the theory of economic policies there is a branch of transport policy.

What is it that distinguishes this branch from all other branches? If one looks in the textbooks[3], the following five peculiarities of the transport sector are emphasized:

1.   The tendency towards ruinous competition;
2.   The tendency towards a structural monopoly;
3.   Competitive distortions;
4.   Irresponsibility of the transporters for the overall economy;
5.   The transport infrastructure as a public good.

With regard to number 1 above, the tendency towards ruinous competition is rooted in what is for this market a tendency toward long-term over-capacity; this necessarily leads to competition and annually drives even efficient concerns from the market. This peculiarity is based upon an apparently atypical cost structure that is determined by low marginal costs and a high portion of fixed costs on the longer life of trucks, ships etc. which makes a 'smooth' adjustment of supply to (sinking) demand impossible.[4] In addition, the peculiarity of this market is rooted in the possibility that demand may decline in a manner that cannot be foreseen by the transporter. Similarly, the economic crisis of 1929 was for many governments the reason for and the beginning of regulation measures in this area; of course we hope that 1929 will not become a routine problem! Regarding number 2 above, a tendency toward a structural monopoly—a theme that is once again entering the economic policy discussions in West Germany in connection with the new technologies (especially information technology)—arises with the so-called supplementary cost functions that is with falling average costs all demand can be satisfied most effectively by *one* supplier. An example of this are network and pipeline associations in the sectors of telephone, electricity and water etc.

This is not the place to describe the 'marginal cost controversy' of the 1940s or the arguments of the anti-marginalists against price policy differentiation (split tariffs, price differentiation, Ramsey prices, peak load pricing). We should, of course, not forget about this cluster of problems, especially for West Germany—the argument of 'raisin picking' that is based on the prohibition of certain business sectors.

With reference to number 3 above, a peculiarity of the transport sector is that competitive distortions between individual suppliers arise. Here our interest is only in the aspects of this broad sector that touch upon the long-distance transport of goods, specifically, the competition between road and rail transport. The problems are:

- Some of the costs borne by the general population such as air pollution, noise etc;
- In general, there are costs that must be borne by road transport that are not attributable to individuals in addition to the loss of time due to congestion (see the work of Pigou);
- Cost distortions as a result of external effects, for example, non-internalized social costs, present a serious and difficult problem in this field. The proposals for solution are either: to set aside the causes of the cost distortions, that is to remove the causes through prohibition or through financial charges according to the principle of origin; or according to the second-best principle, to remove these distortions through subsidy of the allocation distortions (a principle that is constantly asserted in EC integration and harmonization efforts).

As for number 4 above, the transporters have to pursue goals consonant with the entire economy which are not necessarily identical with their own individual economic interests, especially regional or geo-political goals:[5]

- growth policy goals;
- environmental protection;
- energy saving;
- reduction of the danger of accidents.

This brief summary of the special problems of the transport field—in this case, the transport of goods—will not be concluded with a judgment about economic policy (or even an ideological estimate); it will only explain the rationale behind the West German system of industrial long-distance transport and indicate the series of problems to which the deregulation discussion has led and offer a catalogue for judgment which allows us to enter the welfare economics discussion.

The development of long-distance highway transport of goods

A competitive situation between rail, water transport and trucking arose on purely technical grounds, initially in the inter-war period. The history in the Western countries since World War I shows,[6] that the serious regulatory intervention following the world depression affected ocean transport as well as long-distance road transport; the laws after 1948 in their general outlines were continuations of the laws of the thirties.

In the 1920s as transport trucks arrived on the German market (also as a result of war production) and in a time of high unemployment, it was possible with a relatively small expenditure of capital for an individual to purchase a transport truck and to become an independent transporter. Many took this opportunity since the high monopolistic tariffs of the railroad offered a strong incentive. The efforts of the railroad to force the competition out of the market through reduction of tariffs miscarried, and therefore, in 1931 a licence requirement for long-distance truck transport (with the exception of factory transport) was introduced.[7] In addition, the granting of licences was made dependent upon the current need for transport, minimum prices were prescribed and a compulsory cartel was created.

These measures have in essence remained in effect until today, that is after World War II the principles of concessions, quotas, the prohibition of coastal trade, and governmentally approved compulsory tariffs remained. Along with the 'classical' reasons for these regulatory measures—peculiarities of the transport sector, protection of the railroads—there were stronger regional policy reasons that were brought into play, above all the integration of peripheral areas in the British, French and American zones (it should not be forgotten that as a result of the 'iron curtain', the classical land routes to Poland, the Soviet Union, the German Democratic Republic and Czechoslovakia were cut off from one day to the next and became peripherally instead of centrally located).[8]

Transport reforms that were introduced at the beginning of the 1960s (Small Trade Reform of 1961) or that remained stuck in the planning stages (Leber-Plan 1968–72) have in practice brought little change, only a loosening up of the regulations and the management structures. The transport minister can no longer prescribe tariffs, but he does have sufficient considerable influence in matters that are not related to either economic concepts or concepts that are formulated in a juristically clear manner, such as 'the general good', 'the economic relations of transporters', the prevention of 'unfair handicaps' for agriculture, the middle class and underdeveloped regions.

In the 1970s, transport policy discussion focused primarily on individual and public transport of people. The reasons for this were the oil crisis of 1973, a deeply growing interest in ecological questions (the 'dying forest' very much engaged the German conscience) and problems of deflecting legal objections to necessary road construction.

The fact that a discussion about liberalization is taking place now in West Germany is the result of three factors:

1. The collapse of all attempts to revitalize the federal railroad system, which is on the verge of collapse because of its insolvency.
2. The appearance of deregulation measures, above all in the United States,

such as the Airline Deregulation Act of 1978, the Railroad Transportation Policy Act, the Staggers Rail Act and the Motor Carrier Act of 1980 which were aimed at intensifying competition. American deregulation prompted theoretical discussions that had their impact in West Germany[9]; hence, since the end of the seventies, deregulation is 'in' in economic policy discussions—the effect of Margaret Thatcher's policies—the theoretical arguments for deregulation became stimulants in Europe.

3. The third decisive and conclusive impetus was the result of an inactivity complaint in the European Court of Justice and its consequences.[10]

In 1983, the European Parliament brought an inactivity complaint at the European Court of Justice against the Ministerial Council. According to Article 3 of the EEC treaty, common transport policy belongs to the prescribed activities necessary to achieve a common internal market and to achieve the step-by-step convergence of the economic policies of the member states. The European Parliament, however, had been unsuccessful in its long effort to push the Ministerial Council into negotiation although the Commission repeatedly had made proposals in this regard. The Ministerial Council countered that the objective difficulties—the formation of the relationships between the railroads and other transporters in the individual member countries on the one hand and the necessity of utilizing the competitive conditions between countries on the other—had made a regulation or harmonization of economic policies impossible. In addition, the Ministerial Council argued that the Parliament was not empowered to raise an inactivity complaint. This position was not accepted by the Court of Justice. Besides remitting the complaint (which in the long run is of great significance: the previously powerless Parliament can now force the Ministerial Council to act!), the Court, of course, conceded the difficulties that the Ministerial Council had in these transport policy questions, but maintained that the Council had neglected the creation of freedom of services in the area of EC transport, although this was required by the treaty. The court thereby confirmed that transport is in the first instance to be considered as a service, and in addition therefore, national restrictions were to be limited step by step.

Thus, in practice, the Ministry *can* further the harmonization of the comparative conditions between the states. It can also verify the role of railroads in the European Transport Union both within and between the states. It *must*, however, assure freedom of service in transport. Headlines did not immediately follow the judgement. However, the March judgment did forcefully alter the transport policy situation in Europe: national market structures were to be harmonized throughout the European community.

The immediate consequences of this judicial decision in West Germany were almost comical. Instead of the interest groups involved making—understandable — complaints (this is not a value judgement: but the interest of the associations were in fact most seriously injured!), the transport operators and operator associations simply attempted to minimize the results of this judgment. Even at the transport forum on 13 November 1985, the interest groups' representatives denigrated the consequences of this judgment. When the West German transport minister, at the time Dollinger, described the consequences, he aroused plain resentment because no one believed him! When, however, the Ministerial Council decided one day later in Brussels to carry through the liberalization of freight across frontiers including freedom of service, (since postponed to 1993 under pressure from the West German government), they unleashed panic in the transport industry of the Federal Republic. They also demonstrated an apparent failure of policy.

An article, 'The International Transport of Goods in Europe'[11] that appeared in one of the most reputable German scholarly journals on transport policy, *Beitragen aus dem Institut für Verkehrswissenschaft an der Universität Münster*, edited by Professor H. St. Seidenfus, confirmed in four individual contributions, the absolute policy failure in the direction of an EC policy. West German policy, in its dual aims, protection of the railroad and protection of the overwhelmingly middle-class truckers[12], through a policy of postponement, foot-dragging but also repression and trusting in the inaction of the EC Parliament (no one had thought about the Court of Justice) arrived at a situation in which it was forced to act. At the same time, the petitioning interest-group associations in Bonn as well as in Brussels were confronted with the fact that while the fundamental basis of harmonization had been thought through, there had been no concrete considerations about the quantitative or the qualitative impact. That is, the relevant authorities were aware that in 1992 in the EC 'some kind' of consequences would occur, but they could not foresee the practical effects and were of the opinion that 'somehow' all of this would 'regulate itself' even if in an individual sector problems of adjustment were to occur.

In the press too, the theme of EC-92 aroused attention rather belatedly: in the (interest group) association press beginning in 1987,[12] in the economics press beginning in 1988,[13] and in the daily press beginning in 1988. Looking at the chief arguments together, it appears initially that, perhaps not astoundingly in scholarly circles, the five points introduced above which distinguished the trade sector from other service sectors were no longer discussed, that is since the court had determined that the transport sector is a 'normal' service sector like any other and does not have a *sui generis* character. This appeared to be acceptable at least to the economists.

The arguments in general took place at two levels: (1) comparison with the deregulation measures in the United States in the beginning of the eighties (the deregulation measures in the United Kingdom did not 'fit' the comparison as well);[14] (2) the dissection of the comparative differences in the transport sector in Europe (surprisingly—or perhaps not surprisingly if you consider the significant function of transport—this is the first and only service sector that has been precisely analysed).

It would be difficult in this context to describe the position of the West German government and especially the Minister of Transport. In all areas of discussion, but especially in the description of the Federal Government, there is nothing definitive to report, but only a process: not: 'work' but 'discussion in progress'. In so far as this relates to anything it is treated under point 2; here as with other points for the most part the daily press is cited.

With regard to 1, as a typical West German view of the American experience with its deregulation measures one may cite the report of Deputy R. Haungs who travelled to the United States with the Federal Trade Committee of the German Bundestag in 1988 and drew an entirely positive conclusion[15] as the result of his discussions with the regular carrier conference of the department of transportation and the Interstate Commerce Commission: namely that consumption rose by $37 billion, the number of businesses increased from 15,000 to 50,000, 200,000 new jobs were created in long-distance transportation alone and that the sharpened competition in no way created ruinous competition.

Industry, represented by the DIHT (the German Industrial and Trade Association), hopes that liberalization of transports will bring about declining freight prices, and therefore supports pressure on the Federal government to follow the American example in this respect. The other interpretation appears to a certain extent in the view of T. Steinmark,[16] who sees sinking prices on the main routes in truckload transports but rising prices on the lesser routes (that is, negative regional policy effects even with welfare aspects), increased empty transports with rising capacity in trans-border transports and deterioration of the cost–income relationship leading to a tendency toward concentration (ten large concerns move 60 per cent of the goods and earn 90 per cent of the profit). Above all it is the discussion about traffic security which is the subject that has aroused wide interest. Of course, as a result of recent experiences, air travel is central to this interest. In an interview with the German news weekly *Spiegel*[17], the cockpit President attributed falling security standards directly to the deregulation measures of the Carter administration and the resulting pressure on costs. *Güterverkehr* cited with satisfaction the report of the 'Coalition for Sound General Freight Trucking', the judgment of the California court (that demanded regulation measures) and the NBC (National Broadcasting Company) neologism of 'killer-trucks'.[18]

A second problem area that has been discussed by the Commission, the economists, the West German government and the interest-group associations is the comparative differences in long-distance road transport among the EC nations. Three extensive research projects commissioned by the federal minister for transport between the middle of 1987 and the middle of 1988 on the question of competitive differences and the possibility of abolishing these differences (two of these from private research institutions),[19] are actually somewhat late, if you consider the time frame and also that the government ought first to inform itself and then make a decision! Two of the expert studies started from two different scenarios. Scenario A starts with the premise of a complete liberalization of transport of goods, while scenario B takes as its starting point only the liberalization of trans-border transport. The results of the first position (based above all on the experience in the United States) was a recommendation for a complete liberalization policy for general reasons of competition. It was seen as necessary, however, to complete harmonization measures; the impact on the environment and on transport were viewed as less important. On the other hand, in both scenarios the results included the potential for decline in freight, the modal split effects that pointed distinctly to a displacement of freight from trains to roads which would have major consequences for the profits of the railroads; also there would be a negative impact upon the combined transport.

Competition distortions in the EC transport sector and the harmonization proposals of the Commission

The report that dealt with the differences in competitive conditions in six of the EC countries was decisive for the subsequent actions of the West German government since it provided the framework for negotiation possibilities after the deadline of 1 January 1993 was fixed. In a rough sketch, it is possible to divide the state regulation of international commercial goods transport into four sectors:[20]

(a)   taxes, other fees, subsidies;
(b)   market regulations and market oversight;
(c)   social prescriptions and their supervision; and
(d)   technical requirements and their controls.

With regard to (a), the motor vehicle tax ranges between 9,100 DM in West German and 1000 DM in Italy per year (with prototypical allowance for special regulations). Petrol varies prototypical consideration of the territorial principle in cost between 0.32 DM per litre of diesel in West Germany and 0.15 DM per litre in Denmark. It is more difficult to get figures for infrastructural charges (that are levied by France, Austria, Italy and Switzerland) since these charges have different impact according to the primary transport flow. With certain assumptions based on an annual figure of 115,000 kilometres per vehicle we come up with cost differentials up to 13,000 DM per year (the West German fleet comes out somewhere in the middle). These cost differentials have a direct impact upon the profitability of concerns (Dutch transport concerns have the lowest liabilities).

The Commission in Brussels estimates [21] the proportion of petrol and vehicle tax as 4-10 per cent of costs in trans-border transport, motor vehicle cost at 1-5 per cent and motor tolls at 4 per cent. The Commission itself estimates that the cost differential lies within the order of latitude of gross income; the West German transport business realizes therein distinct competitive disadvantages *vis-à-vis* The Netherlands for example (between about 10,000 and 22,500 per truck and year). It is also actually affected doubly: for apart from the high West German motor vehicle and petrol fees for the free West German road system, they also pay, under the territorial principle, for a portion of foreign road construction.

With regard to (c), standards of supervision are controlled by the individual countries; different compensation requirements and different procedures for carrying out the EC social requirements are directly cost-effected. The high West German level of controls has a strong negative impact upon profits. Technical requirements (from the point of view of safety or the environment), such as the continuing control of transport safety (in this regard the German TUV surely is pioneering for all of Europe!), are difficult to estimate with regard to their cost impact, but surely they are highly cost-effected. The daily entry of 28,000 foreign vehicles into West Germany presents a serious factor in transportation security policy.

We, therefore, have a situation with regard to policy in which the following interests face each other in a power triangle.

Long distance transport interests

One of the leading German association representatives, H.E. Kammerer, [22] as the Commission confirmed, points to the negative competitive distortions upon the West German transport business and warns that with the existence of about 30,000 long-distance transport licenses in the Federal Republic the danger arises that with an over-capacity of about 50,000 trucks in the other EC states, the majority of these could press into the German market (which, in fact, with 40 per cent of the transport business, is the largest transport market) and thereby, as economic witnesses confirm, threaten traumatic invasion for railroad and combined transport. His phrase '. . . recognize reality whether we like it or not'[23] stimulated hectic action in the associations as well as in the Ministry. *The interest of the German Minister of Transport in responding to the needs of labour and industry.* Shaken, the minister saw himself in a pressured situation. He saw the problem of over two million unemployed at a difficult time for the governing party, during an election campaign, when one of the chief groups of voters in his party are voters from the middle class (and almost 200,000 people work in the exclusively middle-class organized highway transport business). He had to admit that EC-92 demanded above all a sacrifice from the middle class. He saw himself under pressure from German industry which wanted lower freight rates and pressure from the other EC countries who would not tolerate further hesitation from the German side and who, after their experience with the EC Agricultural Market Organization, were not ready to introduce any highly specific special regulations for further partial sectors.

So far, two measures have occurred to the West German Minister of Transport Warnke. On the one hand, the route was taken to move transport taxes from the national to the territorial principle. Trucks that utilize the West German road system (that is, foreign and West German) beginning on 1 January 1990, will be charged a heavy transport tax of between 2,000 and 7,000 DM. At the same time, the vehicle tax, which is only paid by West Germans, will be reduced by the same amount. This leads to the situation that the high tariffs in Germany will be further affected and the competitive advantage, especially for the Dutch transport business (which overwhelmingly uses the German road system) will disappear.[24] The attempt to maintain freight tariffs at a high level also has the explicit goal[25] of helping the federal railroad to maintain its proportion of freight transport (also the railroad will be subsidized on the costs side). *The interest of the EC Commission in fulfilling the objectives of the White Paper.* The Commission also saw itself forced to act. The three general ideas of the White Paper of the EC for 'completion of the internal market'—creation of one internal market, growth for this market, flexibility of this market—should be realized. This means that comparative cost advantages ought to be borne, 'artificial' competitive advantages separated, and 'fairer and freer' competition should be created. Market organization creation, which is viewed as defective, as in the agricultural, mining and steel markets, ought to be avoided.

From the theoretical standpoint, the truly bold beginning—and also the only possibility that the Commission saw to avoid a new special market organization for the transport sector—therefore exists in carrying 'natural' cost differentials as comparative cost advantages and balancing 'artificial' cost differentials as distortions of competition. Of course, the classification into 'natural' and 'artificial' cost differentials cannot be grounded either in scientific separation nor on the basis of a universally accepted pragmatic catalogue of definitions.

In the highway transport of goods, such things as vehicle taxes, petrol taxes, even tolls, social requirements and their maintenance or requital, technical controls and their supervision can be viewed as 'artificial' differentials. However, the Commission views wage costs, direct taxes, social contributions, and astonishingly, also nationally required measures such as subsidy payments for the procurement of trucks as 'natural' cost differentials. The basis for this is the idea that real cost differentials should be subject to a system of competition through differential taxes and social contribution systems, that is behind this there is a competitive theory orientated to systems rather than to allocation![26] The 'counter-strategy' of the Commission consists therefore in proposals for the equalization of 'artificial' cost differentials in order to exert thereby a calming influence on the national and above all the German counter-strategies. To make a prediction about how the common transport market in 1993 will look would be a senseless undertaking at this time. All discussion today describes the situation to date; the next plan of the governmental commission could be in tomorrow's papers and make much of it irrelevant.

Interesting individual observations with regard to the growing re-orientation of German carriers to special transport with (expensive) special vehicles, an accelerated growth of subcontracting sectors and the elimination of over-capacity are now possible. From the standpoint of theory the interesting thing is, however, that the classical special position of transport policy appears to have been shattered and is now viewed as a service sector 'like all the others', and the EC Commission is no longer prepared to create new 'special policy' as was done for agriculture, mining and steel. At the same time the Commission has arrived, without a proper foundation, at a concept of system competition beyond the concepts of 'natural' and 'artificial' cost differentials for which there is surely no historical precedent and no practical possibilities for comparison (the example of the United States cannot be cited in this regard). The title of this chapter, 'the integration of systems and non-systems', is explained as follows: on 1 January 1993, different transport systems will be integrated in a non-system concept of centralized competition between countries with 'natural', but without 'artificial' cost differentials which is entirely new. Whatever consequences it will have are very difficult to predict. At the beginning of the chapter, we compared the anxiety of the Europeans with the tears of the bride on the morning of her wedding day. In view of this sad course of many marriages—the divorce statistics are distinct symbols—the bride's tears are perhaps ominously justified.

The 'philosophy' of the harmonization in the service sector's international aspects

In the goods sector, the Commission and the Ministerial Council were practically forced to a 'country of origin philosophy' by the Luxembourg judgments (Crème de Cassis case with regard to the German purity regulations for beer, in April 1989, on the utilization of milk products: that is if a special product is produced by a Frenchman in France according to French law, which he is allowed to sell, he may also sell it in Germany; 'sharper' security requirements are to be viewed as protectionism). This is at the moment an internal EC approach which in the long run will make it difficult for either EC or GATT negotiations to continue to argue for this kind of restriction as was done in connection with the import of meat from hormone-treated cattle from the United States.

In the service sector, it is not possible to arrive at an 'integration philosophy' with such clarity. Within the EC, it is possible for the old Commission and the new one to follow the same policy lines utilized in the first structural proposal for the initial sector (transport): no creation of 'new EC policy' but liberalization and beyond this the equalization of 'artificial' cost differentials. It is not possible to foresee what the external strategy of the EC in the service sector will be for the new Commission that came into office at the beginning of 1989. The old Commission had determined that the highest principle for entrance into the EC banking market is the principle of reciprocity—in the interval, no regulations in the service sector have been considered. The proposal for the so-called 'second bank guideline' also begins from this principle and, should the new Commission maintain this principle of reciprocity, it will be equally applied in other service sectors such as transport, telecommunications and insurance.

Notes
1. It is even more complicated, if high-tech products are considered. Is the development of a particular calculator to be ascribed to the production sector rather than the service sector, and is the development of software really a part of the service sector?
2. Estimates, now even from official Soviet sources, indicate that because of the lack of transport cooling and storing services, approximately 30–40 per cent of Soviet agricultural production is lost.
3. For example, some of the most popular textbooks in West Germany and Great Britain which are common in these 5 points are, van Suntum, *Verkehrspolitik*, Munich, 1986; St. Glaister, *Fundamentals of Transport Economics*, Oxford, p. 81; R.E. Just, D.L. Huth, A. Schmitz, *Applied Welfare Economics and Public Policy*, Englewood Cliffs, 1982; E.J. Mishan, *Introduction to Normative Economics*, Oxford, 1981.
4. That is the price equals marginal-cost rule is replaced by the rule: if you offer it, then offer it at the limit of capacity.
5. To study the transcendent significance that transport routes and transport systems can have over the centuries, see: N. Ohler, *Reisen im Mittelalter*, Munich, 1986.
6. According to van Suntum, op. cit., p. 98.
7. Hereafter I will omit all regulations with regard to transport of persons and internal water transport.
8. In the author's homeland it is still possible to view the European continent as a permanent marching region for standing antagonistic armies: across from the Black Forest lies the Vosges on the French side, with the Rhine region in between. The French carry their roads only as far as the ridge of the Vosges, but not below into the

Rhine region. You can bring defences etc. into position here, but the enemy cannot bring heavy *matériel*; and the same applies to the German side. Therefore in this region of a heartland section of Europe there is still no satisfactory east–west trade route, and both regions are still poor.

9. Most important to mention here is A. Friedlander, *The Dilemma of Freight Transportation Regulation*, Washington, 1975.

10. Compare H. Rogge, 'Die verkehrs politische Untätigkeitslage des Europaischen Parlaments: Konsequenzen aus dem Urteil des Europäischen Gerichtshofs vom 22.5.85', *Internationales Verkehrswesen*, 5, 1985, p. 310 ff.

11. H. St. Seidenfus, 'Der internationale Güterverkehr in Europa', appearing in the series, 'Beiträge aus dem Institut fur Verkehrswissenschaft an der Universität Münster, **99**, Gottingen, 1982.

12. See for example the periodical *Güterverkehr*, published in Bonn, which is the association paper of the BDF (Bundesverband des Deutschen Güterfernverkehrs).

13. See for examples the year's output of the well-known German transportation periodical *Internationales Verkehrswesen*, official organ of the Deutschen Verkehrswissenschaft Gesellschaft e.V.—Darmstadt.

14. Compare H. Kirsch, D. Polunsky, 'Die Deregulierung des Strassengüterverkehrs in Grossbritannien' *Internationales Verkehrswesen*, I, 1987, p. 9.

15. According to *Internationales Verkehrswesen*, 4, 1988, p. 229.

16. T. Steinmark, 'Deregulierung zwischen Theorie und Praxis' in *Güterverkehr*, 4, 1987.

17. 'Die Sicherheit wird ausgeholt', *Der Spiegel*, 9.1.89.

18. *Güterverkehr*, 9, 1987, p. 13 ff. and 10, 1987, p. 18 ff.

19. Commissioned by Bundesminister fur Verkehr: (1) 'Ursachen Ausmass und Auswirkungen unterschiedlicher Wettbewerbsbedingungen im europäischen Binnengüterverkehr' Prognos-Institut Basel, August 1987, (FE 90150/86); (2) 'Ordnungspolitische Szenarien zur Verwirklichung eines gemeinsamen europäischen Verkehrsmarktes—Szenarien und ökonomische Wirkungszusammenhange', Gesellschaft zur Forderung der Verkehrswissenschaft an der Universität Münster, May 1988, (FE 90182/86); (3) 'Ordnungspolitische Szenarien zur Verwirklichung eines gemeinsamen europäischen Verkehrsmarktes—Quantitative ökonomische Wirkungsanalyse', Planco Consulting, May 1988, (FE 90207/86).

20. Compare S. Rommerkirchen, 'Der staatliche Einfluss auf die Wettbewerbsbedingungen im internationalen gewerblichen Güterdraftverkehr', in *Deregulierung der Verkehrsmarkte*, series of the Deutschen Verkehrswissenschaftlichen Gesellschaft, e.V., No. B 105/1988, p. 45 ff. Rommerskirchen is one of the authors of the Prognos study. The analysis confines itself to West Germany, Belgium, Denmark, France, Italy, The Netherland; for other countries there were no comparable statistics.

21. K.-H. Schmidt, 'Betrachtungen zu Harmonisierungsdefiziten im Strassengüterverkehr in der EG' in *Deregulierung der Verkehrsmarkte*, p. 115 ff.

22. Compare *Güterverkehr*, 1, 1987, p. 12 ff.

23. Quoted from *Deutsche Verkehrszeitung*, 99, 1987, p. 3.

24. Compare 'Schwerverkehrsabgabe bis zu 7000 DM' in *Süddeutsche Zeitung*, 47, 25–6 February 1989, p. 33.

25. Compare the interview with Minister Warnke with the article 'Schwere Lasten

wieder mehr auf die Schiene', in *Süddeutsche Zeitung*, 51, 2 March 1989, p. 15.
26. Compare K.-H. Schmidt, 'Betrachtungen zu Harmonisierungsdefiziten im Strassengüterverkehr in der EG' in *Deregulierung der Verkehrsmarkte*, from Pamphlet series, various locations, p. 122.

4 The emerging social dimension of Europe 1992

Seamus O'Cleireacain

Europe will never be built if working men and women—white-collar workers, managers, farmers, industrialists, professional people—are not among the first to be involved. [Jacques Delors, President of the Commission of the European Communities, to the European Parliament 17 January 1989]

Introduction

Even before its implementation, the programme to complete the European Community's internal market, popularly referred to as 'Europe 1992' or merely '1992' had become an extraordinary public policy success. It had captured the imagination not only of Europeans, but also of much of the rest of the world. With good reason. The rest of the world has a vital interest in the manner in which European integration evolves. Interpenetration of national economies guarantees that those who shape Western Europe's future also influence, to some degree, the lives of citizens of Europe's trading partners.

Originally established in 1985, the 1992 programme includes a timetable under which intra-European barriers to the mobility of people, capital, goods and services would be removed to yield one integrated internal market comprised of the twelve member–states of the EC. Specifically, the proposals call for the elimination of all restrictions on the movement of workers between member–states, an end to customs and immigration controls within the EC, the removal of all barriers on doing business between the twelve countries and the further development of the EC's trade policy *vis-à-vis* the rest of the world.

While the timetable calls for the proposals to be adopted by 31 December 1992, it is important to note that full implementation will take longer. All barriers are not due to disappear at end–1992, as member–states are often given a further timetable for implementation. At the end of 1992 the EC will still be far from achieving the elimination of barriers necessary to create the unified internal market called for by the Treaty of Rome in 1957.

The EC Commission has estimated that the economic gains of the 1992 programme may add 5 per cent to EC GDP, create as many as two million additional jobs, lower prices by 6 per cent and improve the EC trade balance by 1 per cent of GDP.[1] Approximately half of these potential gains will only occur if European firms become much bigger in size and if there is vigorous anti-trust enforcement to limit monopoly power. Some independent estimates are lower than those issued by the Commission.[2]

With these gains come costs. Very considerable dislocation is expected to accompany implementation of the 1992 proposals. This chapter examines what has come to be known as 'the social dimension of 1992'. The 1992 programme requires measures to cushion social dislocation and to ensure that the standards of social protection attained in member–states are not eroded. The social effects of 1992 include not only the likely effects to be produced by greater mobility of people but also those associated with the other three freedoms of movement: of goods, services and capital. For individuals, the full implementation of the Treaty of Rome promises a lessening of discrimination based on citizenship; this includes discrimination in employment, housing, education and social benefits.

The White Paper

The strategy to complete the internal market was laid out in a Commission White Paper in 1985.[3] The White Paper proposed reliance on a tactic which the Community had successfully used in its earlier institution-building—the coupling of an objective to a staged approach, complete with a timetable by which the objective was to be achieved. The White Paper called for the elimination of three main categories of barriers: *physical* barriers such as customs posts and immigration controls; *technical* barriers such as health, safety and environmental standards, education, apprenticeship and professional qualifications; and *fiscal* barriers such as excise duties and value added taxes (VAT). Barriers would be removed through EC-wide standards, replacing national standards.

Two broad approaches to the establishment of 'Euro-standards' have traditionally been used by the EC. In some instances, member–state national standards have been harmonized by changing them so that they conform to one centralized standard; in other instances, differing national standards have been judged to be functionally equivalent to each other. With a common set of standards in place throughout the Community, the internal market permitting the free movement of goods and services, workers, businesses and capital would be completed.

The Single Act

Legal support for the 1985 White Paper proposals was obtained in 1987, when the parliaments of the twelve countries passed an identical piece of legislation, the Single European Act. The Act called for the Community to 'adopt measures with the aim of progressively establishing the internal market over a period expiring on 31 December 1992'. To implement the Act, the Commission has drawn up a list of 279 proposals which need adoption if the timetable is to be met. The number of proposals will change through time as additional ones are introduced or as some extant are combined together. By mid–1989, the Commission had tabled almost 90 per cent and the Council had adopted a little over 50 per cent of the measures.[4] Progress in obtaining Council approval is uneven during the year, with a rush of decisions usually coinciding with the approaching end of a six-month term of Council Presidency.

On the constitutional front, the Single European Act made several amendments to the Treaty of Rome. In a significant weakening of national sovereignty, the Act strengthened the decisionmaking powers of the Council of Ministers by applying a qualified majority voting system to most matters relating to completion of the internal market. The EC's qualified voting system is a weighted voting system precluding a veto on the part of any one member–state.[5] However, these provisions will not supersede the so–called 1966 Luxembourg Compromise under which member–states agree not to force through an issue deemed by any member–state to threaten its vital national interests. The Act also strengthened the role of the European Parliament by giving it a greater say in proposals before the European Council. The Council's decisions are now taken in cooperation with the Parliament.

The Act also amended the Treaty of Rome by incorporating provisions on industrial relations. A new Article 118a to the Treaty, dealing with the work environment and worker health and safety, requires member–states to 'set as their objective the harmonization of conditions in this area, while maintaining the improvements made'. Directives to achieve the objective would be prepared by the Commission and approved by the Council. In a further amendment, Article 118b called on the Commission 'to develop the dialogue between management and labour at European level which could, if the two sides consider it desirable, lead to relations based on agreement'.

Dislocation

Very considerable dislocation is expected to accompany implementation of the 1992 proposals. In the words of one British industrialist, Sir John Harvey Jones, formerly chairman of the chemical multinational ICI, 'in the next ten years, probably half of all the companies in Europe will disappear or form part of different groupings'.[6] Small and medium-size enterprises will be particularly affected; multinationals already have European-wide operations. Deregulation of road haulage and air transport will open up national markets to competition from other EC countries. European-wide bidding on government contracts will alter public procurement procedures, removing a key component of national industrial support policies. A large number of industries will experience plant closures as industry seeks economies of scale from the larger market. A wave of mergers and acquisitions to get ready for 1992 by creating European-wide corporations has already begun. De Jong reports a virtual doubling of the number of large mergers and take-overs between 1984 and 1985, from 186 to 275.[7] The pace has quickened since.

Although the 1992 programme is estimated eventually to yield net job creation of about two million jobs, one short-term consequence will be an increase in redundancies and a likely rise in the long-term unemployed. There are presently over thirteen million unemployed in the EC, where the average unemployment in 1988 was 11.3 per cent.[8] Although declining, EC unemployment has been in double digits in every year since 1983. About 30 per cent of the unemployed have been out of work for over two years. The EC Commission recognizes that unemployment will remain high for a number of years to come, making it essential that job creation programmes be maintained and that other social programmes be stepped up even as the 1992 programme proceeds. The Single Act noted that some regions of the Community might need more time to implement the internal market and amended the Rome Treaty to require the Commission, under a new Article 8c of the Treaty, to take this into account when designing its proposals.

Gradually, the social dimensions of the 1992 plan are being elaborated. A charter of fundamental social rights was adapted by the European Council at the December 1989 Strasbourg Summit. The need for an EC social industrial relations policy as part of the 1992 programme was recognized from the beginning and incorporated into the Single Act. As EC Commission President Jacques Delors pointed out in 1988 to the British Trades Union Congress, there must be full and broad consultation with those involved in the production of wealth.[9] Pointing to the existence of 'similar mechanisms of social solidarity, of protection of the weakest, and of collective bargaining' throughout Western Europe, the Commission president announced that 'it would be unacceptable for Europe to become a source of social regression'.[10]

The broad outlines of the Commission's approach were expressed in a 1988 interim report of a Commission interdepartmental committee on the social dimension of the internal market and in a September 1988 Commission working paper. The interim report developed proposals in the three areas of freedom of movement of persons; coping with social changes to be produced by completion of the internal market; and the creation of a European system of industrial relations.[11] The working paper identified three related components to the social dimension of 1992 consisting of social policy measures to remove existing barriers to the freedom of movement and of establishment; active measures to encourage new types of labour mobility, (particularly for highly skilled workers); and 'damage-containment' measures to identify and cushion excessive dislocations produced by completion of the internal market.[12] The working paper also produced an inventory of the measures needed. Of the eight listed measures required to implement the social dimension of 1992, only four had been adopted by the Council of Ministers by September 1988.[13]

The working paper also addressed the social implications of the loss of perhaps 250,000 jobs as coincident with the creation of up to two million additional jobs as estimated by the Cecchini Report. It called for the speeding up of actions which would create new employment and the delaying of measures which would destroy jobs. It also drew attention to the escape clause which the Single Act specifically built into the Treaty of Rome through Article 8c which not only calls on the Commission to pay attention to regional disparities in the burden sharing of dislocation when designing directives, but also holds out the prospect of temporary suspensions.

Labour Market Issues

A wide variety of labour market issues will require resolution if there is to be one unified labour market, with workers free to take jobs anywhere within the EC, as called for under the Treaty of Rome's clause on the free movement of persons. These include such questions as portability of pension rights, unemployment compensation, access to social welfare systems, acceptance of educational and professional credentials, trade-union representation, the applicability of labour contract provisions negotiated in one member–state to other member–states, etc.

The further development of the other three freedoms of the Rome Treaty, that is freedom of movement of goods, services and capital will also have major impacts on labour. Among the issues requiring resolution here are those of national differences in plant-closing legislation, anti-discrimination legislation, the transnational applicability of collective bargaining agreements etc.

Labour mobility

The right of European workers to move freely across national borders and to accept any offers of employment is enshrined in Article 48 of the Treaty of Rome. The right is not absolute. Among the exceptions is Article 48.4's exclusion of employment in the public service. Other exceptions to the prohibition of discrimination based on nationality include Article 48.3's and Article 56.1's recognition of member–states' rights to restrict the freedom of movement of persons on grounds of public policy, public security or public health.

The blanket exclusion of public-sector mobility has been narrowed, through decisions of the European Court of Justice, to jobs involving 'the exercise of public authority' or 'the safeguarding of the general interests of the State'. [14] As a result, jobs in the public sector are opening up to workers from other member–states. Further, since December 1987, the EC Commission has targeted four public-sector areas in which it seeks to ensure compliance with the Treaty by ending the practice of some member–states of limiting employment in these sectors to its own nationals. The targeted sectors are teaching, public health services, non-military research and public agencies run on a commercial basis such as airlines, shipping, radio and television, gas and electricity utilities, posts and telecommunications, and public transport.

A list of cases, deemed incompatible with the Treaty of Rome's right of movement of persons, is instructive as an indication of how national laws and regulations are changing to enforce workers' rights. In Belgium, Belgian mineworkers had been given additional leave to complete military service while similar leave was not provided to other EC workers completing military service in their country of citizenship. Belgium had also levied foreign student fees on the families of non-Belgian EEC workers resident in Belgium. In France, non-French EEC workers were denied employment as seamen, and permanent employment as nurses in public hospitals was restricted to French nationals. In Italy, non-Italian EC workers were ineligible to buy subsidized housing and in some regions, only Italian nationals could be licensed as ski-instructors. In the United Kingdom workers and their children were ineligible to receive British university grants. In West Germany, the non-EC spouse of an EC worker could not practice medicine. Most of these instances of discrimination are being eliminated as member–states amend their practices and laws.

A number of EC measures on labour mobility predate the specific 1992 directives drawn up to implement the 1985 White paper. [15] Among the measures already adopted by the Council have been ones establishing equivalence between national vocational training qualifications among job applicants from different member–states, mutual recognition of diplomas in architecture and in the medical and paramedical professions for doctors, dentists, nurses, midwives, pharmacists and veterinary surgeons. Actions to establish equivalence in vocational training qualifications have been concentrated on the hotel and catering trades, auto-mechanics, construction, electrical, agriculture and forestry, and textile and clothing. Implementation is not expected until 1996. [16]

The Council of Ministers has recently approved a directive for a system of mutual recognition of university degrees based on the principle of mutual confidence. However, an escape clause permits member states to use aptitude tests or probationary periods in instances where there are substantial differences in higher-education standards. As mentioned earlier, the 1992 deadline applies to Community decisions, not member–state implementation. For example, in the case of the 1986 Council adoption of the directive on doctors' diplomas, the deadline for full implementation is 1 January 1995.

Portability of benefits

Portability of social welfare benefits is an important component of the social dimension of 1992. Eligibility for some benefits is based on residence. These include education, health, housing. Gradually, through Council or Court decisions, the eligibility of all European workers to these benefits is being established irrespective of the member–state of their origin. Other benefits, such as pensions and unemployment compensation, may be based on either country of residence or country of employment.

Ten years ago, in an earlier effort to make labour markets work better in an age of high unemployment, the Commission had submitted a 1979 directive to the Council permitting the portability of unemployment benefits and early retirement benefits from state to state. The proposals, which had languished on the Council's table, are now part of the 1992 programme, subject to fast-track, qualified-majority voting procedures. At present, unemployed workers cannot migrate from the country where they were last employed without forfeiting their unemployment benefits in that country. In effect, member–states interpret the Rome Treaty's freedom of movement of persons as providing a right to residence only to the gainfully employed. The future portability of unemployment benefits will not mean that workers unemployed in a member–state providing low unemployment benefits are eligible for higher benefits by establishing residence in another member–state. It will merely require that eligibility for benefits not be lost just because a worker changes residence from the member–state of last employment.

Pension portability also presents difficulties. Many member–states have separate social security schemes for public-sector employees. Workers covered by these schemes are currently excluded from the EC legal provisions on social security, limiting the movement of such public-sector workers as teachers and researchers. The Commission proposes to amend EC social security law to permit portability of these benefits.

The fears of migration

For many, the fuller implementation of the Treaty of Rome raises the spectre of massive waves of low-wage Greek, Irish, Portuguese and Spanish workers flooding into the high-wage centres of Europe's core or, alternatively, of footloose capital hollowing out this core by migrating to the low-wage periphery. Political developments in Eastern Europe have extended these fears to include both Eastward capital flows and Westward labour flows. In the past, there has been little empirical evidence to support the propositions of massive core-periphery exchanges of capital or labour. Europe's peripheral areas have not become full-employment magnets attracting massive waves of direct private investment from the EC's centre. Neither are they expected to become the origin of a new wave of labour migration. The increased regional disparities which have occurred within one member–state, the United Kingdom, in the past ten years provide ample evidence that waves of labour migration out of the periphery or waves of capital flows out of the core are unlikely occurrences. The highly publicized investment boom in Spain may be seen as a one-shot phenomenon attributable to economic restructuring and EC entry. It represents a very small share of total EC investment and gross job creation.

The Commission does not foresee any great increase in the migration of unskilled labour. The migration of the 1980s, and that to be expected in the future, differs sharply from the large migrations of unskilled workers which occurred prior to the first oil-shock in 1973. According to the Commission, present migration flows contain a higher proportion of skilled workers and professionals. The Commission's 1988 working paper stressed the need to remove barriers limiting the mobility of Europe's scientific and technical workers.[17]

Nevertheless, the fear of social dumping is pervasive. Social dumping occurs if regional disparities in wages and working conditions drive down standards in high-standard countries, or cause them to lose market share. The Commission's Social Dimension working party identified specific labour-intensive sectors, using largely unskilled labour, as sectors where social dumping could occur. (for example, the construction industry).[18] A large share of construction activity occurs under public procurement. As part of 1992, the previously closed national public procurement systems are to be opened up to EC–wide bidding. Not only will this remove the system of nationally designated suppliers enjoying the monopoly position conferred by the present 'buy-British' or 'buy-Italian' etc. arrangements, it may also produce roving gangs of construction workers, moved by European-wide construction companies from one work-site to another. Unless controlled, successful contractors could recruit workers in the peripheral regions of the Community and move them from one building site to another, ignoring national boundaries, in a manner similar to the use of imported Korean labour on Gulf-State construction projects in recent years.

Job migration

Social dumping may also stem from illicit work done in the black economy to avoid tax, labour and safety laws or the provisions of collective bargaining agreements. The Commission reports that illicit employment is estimated to represent 6–8 per cent of employment in northern member–states, but 10–20 per cent in southern member–states. These differences are a further social dumping threat if they produce capital and job migration. Employers who pay workers 'off the books' obtain a number of cost advantages over their competitors. These include escaping the investment costs required to meet health and safety standards and evading employer contributions to national health, unemployment and pension schemes. The Commission's working party suggested some general measures to control social dumping, but the problem may be expected to persist unless there is vigorous enforcement by member–states. [19]

While differences in wage costs between member–states are very considerable, the extent to which these differences would produce a massive relocation of industrial plants after 1992 is likely to be limited. Labour costs are often not the determining factor in corporate plant location decisions. Distance from market, tax treatment, infrastructure, telecommunications facilities and access to skilled workers are all common influences on investment location decisions. A variety of empirical trade studies have shown that the job displacement produced by stronger core–periphery economic links pales in comparison with that produced by technical change. Nevertheless, there is little doubt but that decisions such as Volkswagen's 1987 purchase of a 75 per cent share of Seat and its plan to spend $6 billion modernizing its Spanish operations influence the pattern of wage negotiations in the German car industry.

The sectors where relative labour costs are likely to have the greatest influence in producing a shift in plant location decisions are those which have already been identified—industries which use labour–intensive technologies relying on unskilled workers. The ability of low-wage member–states to lure investment away from the EC core through subsidized industrial development programmes is controlled by Brussels through its regional development policy. In contrast to the beggar-my-neighbour industrial subsidy programmes of individual states in the United States, the EC policy controls competition between member–states through the use of a common regional policy which designates areas in which member–states may provide regional development aid and establishes ceilings on the degree of subsidization. Even in the case of non-regional aid, such as subsidies or grants provided as part of an industrial policy, the EC has limited the power of governments and even forced subsidized companies to repay part of subsidies already received.

Some job migration may also be expected to occur within the member–states comprising the core. Present differences in work rules among core member-states are inducements to firms to relocate in neighbouring countries in geographical proximity to markets in the core. The five-day work week prevalent in many German industries is particularly threatened.[20] Some harmonization in work rules among core member–states may be expected as 1992 approaches. What is presently unclear is the extent to which this harmonization will be negotiated between employers and labour, the extent to which employers will seek to impose lower working conditions, and the extent to which the labour movements in the member–states will engage in concerted actions in response.

One final point: European labour migration is not limited to intra-EC flows. Recent years have seen considerable flows from Eastern Europe into member–states. West Germany, in particular, has had a relatively open policy with respect to these flows. Italy has become host to a large number of African immigrants. Completion of the internal market will require some harmonization of the immigration policies of the member–states with respect to the rest of the world.

Industrial relations

The appearance of relatively free movement of goods, persons, services and capital will ultimately require a European-wide industrial relations' area. The creation of European corporations chartered under a European Companies Statute will also require EC-wide labour relations. In June 1988, the Commission proposed a range of options for worker representation in enterprises established under its European Company Statute. The proposed options were a German-style worker representation on boards of directors, a Franco–Italian model of enterprise committees and a Swedish-style co-management approach of company-specific arrangements. Article 118 of the Treaty of Rome explicitly holds the Commission responsible for the promotion of closer cooperation among member–states on matters of employment, labour law and working conditions; vocational training; social security; occupational safety; the right of association and collective bargaining.

The scope of the EC proposals for 1992 is so broad that they will have major effects on European labour and management. As European firms place their production and distribution operations on a European-wide basis there will be a tendency to seek to do likewise with industrial relations. If European-wide institutional arrangements are not developed, companies are likely to develop in-house industrial relations influenced by differences in national industrial relations' systems.

As change progresses, some of the social elements in the 1992 programme will constitute a rudimentary European industrial relations' system, operating in tandem with the extremely diverse national systems. There is little likelihood of national systems being displaced by one centred in Brussels. The Treaty-based legal competence of the EC in the field of industrial relations is largely limited to anti-discrimination in pay and employment conditions, health and safety, and to matters which may be shown to impede the four freedoms of movement discussed earlier. Indeed, the 1988 Commission working paper argued against harmonization in areas other than health and safety if the harmonization were to result in job loss. [21]

European industrial relations' practices vary enormously from country to country. The rate of unionization varies greatly between member–states, from 70–80 per cent in Belgium and Denmark to 15–30 per cent for France, Spain and Portugal. [22] Labour standards are produced by a mix of law or regulation on one hand and collective bargaining on the other. The mix varies between member–states. In some member–states such as France, there is heavy reliance on centralized national laws and regulations. In others, such as the United Kingdom, there is a greater reliance on collective bargaining. These differences between state intervention and collective bargaining reflect differences in national traditions, legal codes and systems of government. Within the EC, there is a recognition that it is not desirable to replace these differences by a centralized, Brussels-based system. However, there are areas in which some common standards can be applied either through adoption of new standards or through approximation of existing national standards.

In some member–states, collective bargaining agreements reached in the unionized sector are also applied to non-unionized workers if certain conditions are met. In Belgium, agreements signed within the Conseil National du Travail (National Labour Council) may be made binding on all employers by royal decree. [23]

Similar arrangements are applied in France, Ireland, Luxembourg and the Netherlands. Provisions for general applicability of agreements are rarely used in West Germany except in sectors with a large number of firms. No provisions for general applicability exist in the United Kingdom or Denmark, although in the Danish case acts of Parliament have been used to introduce general applicability of agreements.

As mentioned earlier, the EC had been active in the labour field long before the 1992 proposals. EC-based labour standards have taken the form of regulations or directives issued from the Commission or the Council, and decisions handed down by the Court of Justice. The EC's earlier involvement included a series of regulations and directives on freedom of movement of workers and a series of Council directives on labour law. [24] The labour law directives have included the 1975 directive on approximation of member–state laws on collective redundancies, the 1977 directive on workers' rights in transfers of businesses and mergers, the 1980 directive on worker rights in insolvencies and the 1975 and 1976 directives on equal pay and equal treatment for men and women. The 1975 collective redundancies directive is an EC-wide plant-closing law which requires a minimum of thirty days advance notification of mass redundancies.[25]

Not all Commission draft directives have been adopted by the Council. Disputes between member–states have delayed the adoption of a considerable number. Stalled directives include the 1983 draft fifth directive on company law, the draft Vredeling directive on worker rights to information and consultation, and draft directives on part-time and temporary work and on parental leave. Worker representation has been a particular source of dissension among member–states when the Council of Ministers has considered Commission proposals. The Council did not approve the Commission's draft company law directive which contained a clause that one-third of the directors in large enterprise be selected by workers. In November 1988, despite United Kingdom opposition, EC trade ministers agreed to ask the Commission to redraft its proposed European company statute. Attempts to draft a directive establishing a common European law on corporate mergers were also delayed in part because of disagreements on the extent to which there should be labour representation on boards of directors but also because of disagreements on the size of mergers to be subject to prior investigation under EC competition policy and the applicability of the merger policy to state-owned enterprises.

With the Single European Act's addition of Article 118b to the Treaty of Rome, the possibility now exists for some collective bargaining at the European level should the social partners so desire. The Commission's initiatives in this direction, begun with the 1985 Val Duchesse meetings between the ETUC and the employers' federation, UNICE, have not been particularly dramatic, consisting of a number of study-group reports and opinions on the Commission's Annual Economic Report. Examples of European-wide labour–management relations have begun to emerge at the sectoral or company level. These include the 1985 agreement between the European Metalworkers Federation (EMF) and the French multinational, Thompson to establish a five-country works' committee. The EMF has also established cross-border trade-union committees in such companies as Airbus Industrie, Bull and Continental Can.

The EC 1988 interim report of the Commission working party identified three main areas for action on a social dimension to 1992: harmonizing standards in fields such as health and safety and of regulations limiting worker mobility or distorting competition; seeking long-term convergence of member–state social protection schemes; and encouraging innovation and experimentation in social policy.[26] Six draft directives on safety were submitted to the Council in 1988 and specific measures are being drafted for three high-risk industries—fisheries, agriculture and construction.[27]

In early 1989, the Council of Ministers approved the machinery directive which harmonizes national health and safety standards for the design and construction of new machinery. These standards are to be developed by the appropriate EC standards' bodies. The machinery directive is of great significance. It not only replaces a multitude of non-tariff trade barriers by a single set of technical standards for intra-and extra-EC trade in machinery, it also helps establish a common standard for work-place health and safety. Finally, it is an example of a sweeping, European fast-track 'new approach' to harmonization adopted by the European Council in May 1985. Under the approach, harmonization is restricted to essential ingredients rather than all detailed specifications.

The 1988 working party rejected as unrealistic a European collective agreement in the near term because of the present member–state diversity in the rights and obligations of contracting parties. [28] Instead, it proposed the establishment of minimum European-wide worker rights. These were to include the right to be covered by a collective agreement or analogous arrangement; to social security coverage; to prior notification and consultation on important changes affecting the firm; to defined contracts for part-time or flexible work; to a standard employment contract; and to a decision on continuing training. Proposals in these areas are presently being drafted. [29] The European Trade Union Confederation (ETUC) and the European employers' federation UNICE have been establishing the ground-rules for worker notification and consultation so that the Commission may draft a directive on the topic.

In the area of worker fringe benefits, a proposed directive on group life insurance and pension fund insurance was being prepared by the Commission in early 1989. In permitting greater competition in the insurance industry, the proposed directive presents an opportunity to lower European labour costs without affecting the coverage of individual workers. However, the opening up of national markets in this area will threaten any special relationships which have developed between domestic insurance firms and the social partners.

The reaction of labour

Both European labour and management are heavily involved in the design of 1992. At the national level, they have sought to influence member–state position in the Council of Ministers. At the Community level, through ETUC and UNICE, they seek to influence Commission draft proposals and also participate in the Val Duchesse process. While supportive of the 1992 programme, European labour is insisting that it be given adequate representation in the making of decisions which will affect more than a hundred million workers. For labour, there is a need to ensure that any gains from the 1992 experiment are fairly shared between the social partners; that there are adequate safeguards to limit disruptions in employment, wages, and working conditions; and that established social rights in some countries are not eroded by competition from countries lacking those rights.

The fears of social dumping have been discussed earlier. Organized labour seeks to ensure that the high wages and fringe benefits available in some of the highly developed EC countries are not threatened by industry being drawn to low-wage, periphery countries such as Greece, Spain, Portugal and Ireland, and to Italy. On the other hand, industry and labour in these less-developed countries of the EC are concerned that increased competition from the highly developed EC countries will threaten their existing industries.

At the member–state level, organized labour has supported Commission draft directives which would raise national working standards. The British TUC, for example, has drawn attention to the Thatcher government's opposition to a number of draft directives, including those providing added protection to part-time workers and temporary workers, and a directive on parental leave.[30] Organized labour has also developed positions on directives which influence citizens as consumers. In the case of harmonization of VAT rates, the TUC has expressed its opposition to any British government proposal which would raise taxation of low-income taxpayers.[31]

Notes

1. A summary of the Commission's study has been published in the United States: Paolo Cecchini, *The European Challenge 1992: The Benefits of a Single Market*, Gower, Brookfield, Vermont.
2. The econometric forecasting firm Data Resources Inc. estimated that the GDP growth rate would increase by only 0.6 per cent and that there would be an almost negligible impact on unemployment.
3. Commission of the European Communities, *Completing the Internal Market*, 1985.
4. Commission of the European Communities, *Fourth Progress Report of the Commission to the Council and the European Parliament concerning the implementation of the Commission's White Paper on the completion of the Internal market*, Brussels, 30 June 1989, p.6.
5. In the EC's weighted-majority voting arrangements, the four larger member–states (France, German, Italy and the United Kingdom) have ten votes each, Spain eight, Belgium, Greece, the Netherlands and Portugal five each, Denmark and Ireland three, and Luxembourg two. Fifty-four votes constitute a majority although in some cases at least eight members must be included in the fifty–four.
6. Commission of the European Communities, *Eurecom*, New York February 1989, p. 3.
7. Henk Wouter De Jong (ed.), *The Structure of European Industry*, London Kluwer Academic Publishers, 1988, p. 6.
8. Commission of the European Communities, *Annual Economic Report 1988–89*, Statistical Annex, Brussels, 19 October 1988 p. 11.
9. Jacques Delors, '1992: The Social Dimension', address to Trades Union Congress, Bournemouth, 8 September 1988.

16. Commission of the European Communities, *Social Europe—The Social Dimension of the Internal Market*, Brussels p. 29.

17. Commission of the European Communities, *Social Dimension of the Internal Market*, Commission Working Paper Sec (88) 1148 final, Brussels 14 September 1988, p. 18.

18. Commission of the European Communities, *Social Europe—The Social Dimension of the Internal Market*, Brussels 1988 p. 66.

19. 19. Ibid.

20. 'Is the German Featherbed on its Way Out?', *Business Week*, 13 March 1989, p. 62; Ferdinand Protzman, 'Affluent German Unions Fear 1992', *New York Times*, 8 March 1989.

21. Commission of the European Communities, *Social Dimension of the Internal Market*, Commission Working Paper Sec (88) 1148 final, Brussels 14 September 1988, pp. 25–8.

22. ILO data quoted in Commission of the European Communities, *Social Europe – the Social Dimension of the Internal Market*, Brussels, 1988, p. 113.

23. Commission of the European Communities, *Social Europe—The Social Dimension of the Internal Market*, Annexe 11 Brussels 1988.

24. Ibid, Annexes 1, 7.

25. Council Directive 75/129 of 17 February 1975. *Official Journal of the European Communities*, no. L 48/29.

26. Commission of the European Communities, *Social Europe—The Social Dimension of the Internal Market*, Brussels, 1988 pp. 69–70.

27. Commission of the European Communities, *Social Dimension of the Internal Market*, Commission Working Paper Sec (88) 1148 final, 14 September 1988 p. 49.

28. Ibid., p. 73.

29. Ibid., p. 50.

30. Trades Union Congress, *Maximising the Benefits Minimising the Costs: TUC Report on Europe 1992*, London 1988. p. 18.

31. Ibid., p. 13.

Part III The technological challenge

5 Forging the European technology community

Pierre-Henri Laurent

The relaunching of the European movement aiming to complete the one internal market has transpired in less than the last decade. In its bare outlines, it is evident that the European Community (EC) has emerged from the static and non-productive seventies with its Europessimism and Eurosclerosis and moved dynamically toward the challenge of further integration of the diverse national economies via the formation of a single mass economy of scale. Central to this Western European resurgence have been the active and determinant roles of the high-technology industry leadership and the EC Commission pressing for the creation of a globally competitive European technology community. Put succinctly, the major catalyst for the White Paper of 1985 and Single European Act (SEA) of 1986 that propelled the twelve to go to 1992 was the alliance of the European public and private elites promulgating the need to construct a centre-piece in telecommunications electronics aerospace and finally an advanced television centre-piece for an accelerated future European market growth an true global competitiveness.

This chapter asserts that this Berlaymont and highest of the high-tech coalition proceeded to design and then manage, in essence co-design and co-manage, a series of programmes such as ESPRIT and Eureka that focused on swift development in this critical sector of a collaborative multinational electronics research capability. It furthermore documents the emergence of two related strategies and policies which grew out of the research and technology development (R & TD) programmes of 1979–86, and which in tandem with the first point, resulted in a real revolution in the Western European political economy. First, there was the business leaders' belief that the propelling force to market revitalization and competition was more liberalization and deregulation of the economic marketplace. Second, their position, in accord with the Commission, was that there had to be a link between the drive to a collaborative information technology (IT) and electronics technology development and the completion of a single European market, that is a borderless Europe.[1]

This rapid evolution of Eurotech—a community research and technology policy—in the eighties was the result of the business and technocratic awareness that the global third industrial revolution of the seventies necessitated swift and massive economic transformation. Their maxim become 'cooperation should be a corollary of competition'. Not only was there the notion of forging a mega-economy to make European trade competitive with the Japanese (and less so, with the Americans), there was also the strong collective feeling that more Euromarkets and European export trade muscle in the cutting-edge sciences and technologies had to be attained at whatever price. To compete with IBM, DEC, AT and T and Japan's Toshiba, Fujitsu and Nissan, Europe's information technology, biotechnology, robotics and other related electronics-based industries with aviation and space offshoots had to be earmarked as the crux of this fast-forward movement to advance EC member–states and their R & TD. Although the eighties began with an accent on accelerated pure scientific cooperation, the commercial application of innovations was quickly added into the supercritical projects to ensure that the European scientific ideas were translated quickly into market products that could compete. The Europeans had numerous examples of lost opportunities: the most stunning was Philips' 'discovery' of both the VCR and the compact disc, which the Japanese (Sony) had first converted into marketable products.

The actual basis of the political coalition that emerged in the late seventies came together just as a severe technology-markets' revolution hit the Europeans. The technocratic experts of the Commission, employing Etienne Davagnon the Belgian Commissioner as their power base, and the leading representatives of about forty high-tech industries, all of them so-called national champions, recognized that Japan and the United States were pulling ahead swiftly in the global race for electronics-sector markets. Playing the principal and weighty role in this construction of a lobbying alliance were the very large Eurotech enterprises. Although events later in the eighties would increase the involvement and opportunities of small and medium-sized firms in the 1992 venture, the initial strategic impact on both the Commission and national governments came from the two or three biggest technology corporations in each of the Community member–states. Western Europe, which into the very early seventies could claim the lead in some major research and product development, was steadily losing ground by the late seventies, especially in the foremost sector of telecommunications and in the material and life-sciences research areas. This recognition of a technology gap, as reflected in the high-tech trade figures, came slowly, but even when perceived had to fight to overcome the dominant thrust and overarching power of the economic recession and stagnation. Even though the depths of the economic downturn seemed to be in 1979, it was in that year that the first steps were taken to overcome the multiple divisions and overall fragmentation of European industry.

The first lessons learned about the weaknesses of the many autonomous and relatively weak national economies appear to have been both a success story and a failure. The success was the European Space Agency (ESA), and the failure involved the effort to have *one* European mainframe in the UNIDATA project. Technological cooperation on peaceful uses of nuclear energy had been an initial field of EC endeavour with Euratom, but the formation of the ESA in 1973 provided a multi-industry process when two previously low-level space efforts were united in a thirteen state project pressed into being by Paris.[2] Nine of the now twelve of the EC (that is, excluding Luxembourg, Portugal and Greece) plus Norway, Sweden and Switzerland built a consortium in military and civilian space technology to break the American and Soviet monopolies. The ESA was an important European venture because it went beyond the essentially basic research of the older CERN organization and became an organ for commercial development in the highly specialized field of space launchers. The European industrial concerns involved included many of the prime catalysts for the EC programmes of the early 1980s. Similarly the participants in the Unidata project attempted to become the vanguard of the high-tech lift-off programmes in the R & TD realm.

While the parallel airbus and later Eureka operations of multinational collaboration were plainly commercial technology developments with target-orientated research and ultimate product creation, the ESA moved quickly from its inception in 1973 as a centre for pure research into transnational aerospatial ideas to its launcher production phase in the late seventies. The ARIANE launch vehicles, which in the mid-eighties were to benefit from American Challenger disaster, have been an enormously successful mix of French, German, British and Dutch collaboration. Planning for the launching of commercial satellites forged a new and significant Western European lobby force that meshed with sympathetic elements in the Davignon industrial and technological divisions of the Commission, resulting in a most potent and influential political force.[3]

The breakthrough came in a series of highly integrated second-generation EC telecommunications programmes conceived and implemented between 1979 and 1985. The dominant forces in the new technologies firms made their desires clear and compelling to Brussels and the other national capitals. Working in concert with EC staffers, Siemens, AEC and Nixdorf of West Germany; Bull, Thomson and CGE of France, Olivetti and Stet of Italy; General Electric, ICL and Plessey of the United Kingdom, along with Philips of the Netherlands, banded together both to lobby their own governments and bring significant parts of the European banking sector into their overall plans. Their closely related aim was to create viable collaborative programmes with the EC utilizing national, community and products' funding for a technological *relance* but, above all, to do so in a completely *new* European political and economical context. This Eurotech coalition wished to use the EC instrument for these programmes, but equally wanted their own governments to respond to what they viewed as the new macroeconomic conditions of the eighties and nineties. They argued that the best means for a revived European trade competitiveness was not just putting their money mainly on a one-sector expansion and elaborate cross-frontier teamwork, but more integrally, moving the political economies to supply-side initiatives, liberalization and deregulation of the markets.[4]

The origins of ESPRIT, BRITE and RACE in the early eighties were paralleled the emergence of a consensus advocating a shift in the European economic order away from government and towards greater market freedom. This union of forces, with the heavy economic and political clout of the high-tech elite at its centre, developed the ideas of reorientating the basic economic outlooks for Western Europe and confronting the task of completing the European general market.

This strategy was possible between 1982 and 1984 because the initial successes of the ESPRIT programme reflected strong national government approval and support.[5] More extensive synergism between ministries and cabinets, the Commission and the high-tech leaders became visible and influential in the formation of the Roundtable of European Industrialists (a group of forty), the Monnet Committee Number 2 (composed of Europarliamentarians), the Roundtable of European Bankers, the so-called 'Kangaroo group' of European legislators, and even the Schmidt–Giscard monetary group. The Kangaroo group and the Monnet Committee were the prime initiators in the EP of what would become the Single European Act (SEA), that is the move to counter the paralysis of Community decisionmaking via institutional changes in the voting procedures. These lobby associations drove relentlessly to support the formation of a European-wide response to the economic malaise and the resulting political dilemmas. When two new groups representing European labour and consumers joined this amalgamation, the way was cleared for an idea like EC-92.

In the numerous industrial meetings consideration of the bleak European technological future without cross-frontier collaboration began to compete with the transformation of the European political economy. In the first domain, the argument about Japanese and American technological domination, especially in mainframes and computers, was explained in part by acknowledging Europe's inherent disadvantages. These included meagre venture capital which was frequently spread too thin and the duplication or replication of basic research efforts. Most importantly, Europe suffered from a disturbing disconnection between the 'idea stage' in research and the finished product in the market stage. Also the segmented and separated talent pools divided Europe into isolated university and institute teams and the private-sector labs. Considering all these obstacles to efficiency and growth, there was little disagreement that the smallness of even the titan European technology groups, when compared to their world competitors, made the European firms incapable of sustained and successful competition. As national economies divided Europe, it could not compete for its own markets let alone others; the solution had to be industrial and technological cooperation and coordination which would make Europe virtually into one large market. The basic decision of the 1982–4 period was to emphasize the internal priorities of consumer electronics, computers and integrated circuits/currents in the grand design of Eurotech.[6]

The members of the Commission who had created the pilot phase of ESPRIT up to 1983 and then its first actual phase (1983–7) helped to convey the technology-sector thesis and also extended the argument to the 'men across the street', the European Council. This pressing forward of the combined R & TD and economic restructuring design was facilitated in the 1983 and 1984 Council meetings at Stuttgart and Fontainebleau. A broad plan to ensure the relaunch of the Community was tied to a series of conflict resolutions. By the time of the Milan Council summit of 1985, a compromise on a budget rebate for Mrs Thatcher and a series of transitional-stage CAP reductions were negotiated. The end of the decade-long budget squabbles was a trade-off created for the United Kingdom which substantially reduced the Prime Minister's reluctance about the more far-reaching move to spur market completion. Given the popularity in 1984–5 of the draft treaty establishing the European Union conceived by Altiero Spinelli, the member states were willing seriously to discuss reviving the integration vehicle.

The Eurotech alliance was already leaning hard on the governments about specific needs. They pointed out to these national representatives the need to face directly the nagging central problems of technology standards and national public procurement. The debate to influence national leaders about supporting ESPRIT and also the RACE and BRITE programme became fused with the necessity to overcome both the diversity of standards and the rigidity and inefficiency of nationally subsidized procurement policies. In supporting these requests for larger national funds for the first ESPRIT monies, the Roundtable businessmen observed that radical changes in procurement and standards would mean the end of twelve telecom equipment systems and standards with multiple incompatibilities and no economy of size. Comprehending that this would be a touchy and sensitive political direction to take, the business elite indicated that there were important and valuable production and trade benefits. The prime example was harmonizing standards so that interchangeability and connectibility would result and make Europeans more competitive.[7] In effect, the Europtech elite not only asked the Commission to make common cause with them on the issue of focusing on transnational technology precedence, but also asked Berlaymont and the separate governments to follow their lead in the restructuring of delicate issues like procurement and standards.

The entire strategy and actual programmes for uplifting European telecom, aerospace and computer technology in the new era of global competition was by 1984 a part of a broader and more comprehensive private-sector plan to address the reform of the entire economy. It was the Roundtable industrialists, particularly in the two reports authored by the Volvo and Philips chief executives, especially in the Dekker Report, who moved the debate in the European capitals toward the corporate restructuring goal through a revised EC framework. Their desire for a new market orientation that would release these dynamic forces called for government intervention on both the national and Community levels. The operating rules of the fragmented economies, so said these industry leaders, required a rapid, deep and profound alteration to the supply-side orientation, with all the benefits of privatization. The presentation fell on receptive ears in much of Europe, where numerous Socialist coalitions and the Keynesian responses to the economic ills had failed. The ideology of Prime Minister Thatcher and the influence of Reaganism in 1984–5 was not insignificant in gaining support for EC high-tech programmes and backing for the more revolutionary reorientation toward a revived *laissez-faire* Europe and Euromarkets. The point was made that nations had to use the EC to gain more power for the product and capital services and even social aspects of the single-market goal. This would not be increasing the competence of Brussels, but, in effect, a decentralized transfer of power by the states to the more energetic, dynamic and reinvigorated European high-tech collaborations accomplished through a Community-based harmonization of markets into one common domestic market. [8]

There were concrete bridges between the first EC high-tech programmes and the emergence of the 1992 goal in the White Paper. The essential one was the Eureka programme, but the coming of age of the commercial aircraft consortium in the early eighties and the development of a new and different European advanced television programme were instrumental too.

The White Paper and the Single European Act evolved in parallel with the appearance of Eureka. It was not incidental happenstance that the original French scheme of 1985 to respond to the American SDI challenge swiftly gained favour in the Community and many EFTA states dedicated to R & TD programs. In contrast to the prime scientific initiatives of the 1979–84 period, Eureka found favour because it addressed directly the commercial marketing and financing orientation of collaborative advanced-stage projects through fifty-fifty public/private funding. Within a year, at Paris and then Hanover, some sixteen states and the EC Commission joined forces to launch the agency with ten and then seventy-four big-price-tag Eurotech projects. By 1987 the total was 109 accepted civilian proposals valued at over $5 billion, with predominant French involvement (in over half of the projects) and other major participation by the United Kingdom, West Germany and Italy. The priority projects covered every aspect of IT, lasers, robotics, biotechnology, environment and transport protection ideas, with two major endeavours gaining the largest support: artificial intelligence development and third-generation robots and factory automation were chosen as the two cardinal priorities.[9]

Eureka was a momentous stage in the high-tech movement and the commitment to wider and more efficient European markets. Although not exclusively an EC programme, this bottom-up, decentralized endeavour stressed the mixed approach to maximizing European civilian technology prowess in the new future. The promotion of cross-frontier technological cooperation and coordination had resulted in a major European consortium designed to make these advanced technology firms competitive. Once again, in Eureka, the most prominent technology organizations were in league with the Commission.

Out of Eureka and its predecessors came not only the White Paper but also one of collective Europe's main surges to be a high-tech competitor where it would count. Eureka Project 95 was the combined West European programme for attaining a sizeable market share in the future new technology of television. High-definition television (HDTV) for the Europeans had evolved in a context of projected severe competition from the Japanese and their MUSE system. There was therefore both an economic/commercial and a political reason for the one-common-effort approach or purely European standard for HDTV. In the mid-eighties this particular high-tech issue became a means for evaluating an emergent ETC's success as a world competitor, particularly since the new European standard would be adapted to current generation sets.[10]

In May 1986 at Dubrovnic, the Europeans chose to deviate from the proposed Japanese standard which the Americans had implicitly accepted. The European decision to have a European HDTV standard was not born simply out of fear of Japanese domination of television sales, but a reflection of the concord among French (Thomson), Dutch (Philips) and (less so) German (Bosch) and British (Thorn) television manufacturers. Their dialogues resulted in co-action based on their previous research, entitled MAC (Multiplexed Analog Components), and its acceptance as the basic systems approach in further HDTV research and development. Since European HDTV research had differed in its mode of procedure from the Japanese work, the Europeans also believed their distinct route to one enhanced TV transmission and display technique was even more mandatory if they were seeking global TV sales. The EC Commission became a prime advocate of the MAC HDTV model, pushing the Eureka 95 and its first production equipment demonstration in late 1988. The full system, due for completion in 1994–5, became a principal high-tech goal in Europe alongside of 1992. The upgraded role of Philips in this venture has led to a minor schism between the Dutch–French–English alignment and the Germans, resulting in some German hesitancy but not outright opposition. These same firms have staked out their central roles and influential powers, becoming the most rationalized of Europe's technology businesses, the political movers in the Community.[11]

The original impetus for the Commission White Paper of June 1985 came from the private-sector high-tech collaborators of the seventies, mostly those in the ESA and the Unidata experiments. Finding sympathetic elements in the EC Commission, these industries combined first to garner support for massive funding of cross-national advanced-technology programmes housed in the EC. Convinced that this sector needed such support to make European markets grow globally, the business and Commission union, sometimes now called Eurotech, expanded their basic notions in several ways from 1982 to 1985. They moved to attract non-EC high-tech participation. They lobbied their governments to make both a central shift in their political economies via liberalization and to adopt a comprehensive recommitment to completing the single European internal market.

This swift succession of events was to make technology a most favoured area in national and European terms. From this advocacy of European solidarity in one sector, the trade vitality and prosperity of the entire Community in world trade became the subject of dialogues between national, intergovernmental and business leaders. The evolution of European information and telecom enterprises was tied to gaining more markets, to further liberalization of domestic markets and finally to achieving a single market. The route from ESA to ESPRIT, from ARIANE and Airbus to Eureka and HDTV's MAC represented the steps that led to the 1985 White Paper and the SEA. The link between these cooperative institutions and structures, initially based on sector self-interest and larger market share, became the more revolutionary abandonment of all national economic boundaries as outlined in the White Paper. Working together at the highest technical levels, single market completion and reorientation of the western European political economy produced an avalanche of compelling changes and the most important *relance* in the contemporary European integration process since the Treaty of Rome.

References

1. Helen Wallace and Adam Ridley, *Europe: The Challenge of Diversity*, London., Routledge & Kegan Paul, 1987, *passim*; Andrew J. Pierre (ed.), *A High Technology Gap?: Europe, America and Japan*, New York Council of Foreign Relations, 1987, especially the essay of Henri Cusien and Carlo de Benedetti; and Pierre-Henri Laurent, 'The European Technology Community: The Meeting of the Elites and the Completion of the International Market', *Il Politico*, **L11**, no. 2, 1987, pp. 309–16. See the essays of M. Sharp and J. Pinder in Juliet Lodge (ed.), *The European Community and the Challenge of the Future*, London, Pinter, 1989

2. 'Research and Technological Development Policy', *European Documentation*, 2/1988; also see the presentation and arguments of Jacques Pelkmans and Alan Winters, *Europe's Domestic Markets*, London, Routledge & Kegan Paul, 1988; and 'Europe

Without Frontiers—Completing the Internal Market', *European Documentation*, 4/1987
3. Commission of the European Communities, 'Telecommunications: the New Highway for the Single European Market', COM 15/18 October 1988; Michel Richounier, 'Europe's Decline is not Irreversible', *Journal of Common Market Studies, XXII*, no. 3, pp. 227–45. See also *The New York Times*, 22 May 1988
4. Interview, Pierre DeFraigne, 14 November 1985 and 'Knights of the Roundtable: Can They Move Europe Forward Fast Enough?' *International Management*, July 1986, 22–5; Lord Cockfield, 'Completion of the Internal Market', Institute of International Economics (Washington), 24 May 1988 and *Financial Times*, 23 March 1988; *The New York Times*, 22 May 1988.
5. Lynn Mytelka and Michel Delapierre, 'The Alliance Strategies of European Firms in the Information Technology Industry and the Role of ESPRIT', *Journal of Common Market Studies, XXVI* no. 2, 1987, pp 231–53; Pierre-Henri Laurent, 'Renaissance Through Technology: The European Decision on ESPRIT', *The Fletcher Forum*, Winter 1985, pp. 155–67
6. Interviews with numerous members of the Roundtable and the EC Commission staff, especially under the Thorn presidency with Corrado Pirzio-Biroli.
7. Commission of the European Communities, 'Esprit for Europe's Future', Brussels 1984
8. The ESPRIT Programme, *Project Synopses (advanced micro-electronics, software technology, advanced information processing, computer integrated index, and associated sub-programme overviews)*, EC Commission, Brussels 1987; *Esprit 87: Achievements and Impacts*: procedings of the fourth annual ESPRIT Conference, Brussels, 28–9 September 1987
9. On Airbus, see Keith Hayward, 'AIRBUS: Twenty Years of European Collaboration', *International Affairs, LXIV*, Winter, 1987–8, pp 11–26. On Eureka, see 'Declaration of Principles Relating to Eureka', ministerial meeting, Hanover, November 1985; 'Eureka', the M. Secretariat of European Research Agency, Brussels, 1987; the M. Lucas essay in Stephen Gill (ed.), *Atlantic Relations: Beyond the Reagan Era*, New York St. Martins Press, 1989; and also *New York Times*, 28 June 1988
10. Commission of the European Communities, 'Proposal for a Council Directive on the Adoption of Common Technical Specifications of the MAC/Packet Family of Standards for Direct-Satellite Television Broadcasting', COM (86) final 1, Brussels 22 January 1986
11. *The Wall Street Journal*, 23 January 1987 and *High Technology Business*, April 1988. The late entry of the United States in the crucial competition will probably create a competition triangle. See also *Christian Science Monitor*, 1 July 1988 and *The Economist*, 2 July 1988.

6 Technology, competitiveness and cooperation in Europe
Alain Dumont

Eurofailures

The first oil shock initiated an era of justified pessimism in Europe that prevailed until the mid-eighties. If European double-digit inflation and increasing unemployment were more publicized and negatively felt issues at the time than technological decline and diminishing competitiveness, the latter were nevertheless as real and as harmful as the former.

Loss of competitiveness affected a number of activities in which Europe had hitherto enjoyed a comfortable position: German cameras and British motorcycles for example, that had been famous world-wide and global leaders were replaced by Japanese imports. *An incapacity to reach competitiveness* characterized European industry for those new products which have become the commodities of the electronic age: hi-fi, pocket calculators, VCRs, video cameras, etc. There again (with the notable exception of the Swatch success), most battles were lost to the Japanese. These European setbacks were patent for everyone: shopping was the easiest way to realize Europe's ailing competitiveness.

But a less visible and even more serious European decline also originated in these days. To realize it, one had to go to this time to the United States, no longer shopping but visiting laboratories and young Californian high-tech ventures. In 1971 Intel developed the first microprocessor, and in 1973 the first genetic recombination was performed by Cohen and Boyer in Berkeley: two major innovations that led what became information and biotechnologies. These two purely American ingredients are today giving birth to the 'third industrial revolution', and are going to shape the twenty-first century[1].

At the end of the eighteenth century, Europe had been the scene of the first industrial revolution, based on steam engine technology; about one century later, European (and American) technological innovations such as electricity and gasoline or petrol engines had nurtured the second industrial revolution. At the end of the twentieth century Europe, for the first time in its history, might well be a follower and no longer a leader in technology.

Even worse, it might come third in the 'triadic' race[2]. The United States Office of Technological Assessment declared in 1982, with regard to the global competition in integrated circuits: 'the battle for leadership in this industry is a battle fought between American and Japanese producers on US and European markets'. Indeed, between 1980 and 1985, Europe's share in the global production of electronic equipment fell from 26 to 21 per cent, covering only 88 per cent of its own domestic market. In 1984 the same OTA stated: 'Japan will be our most serious competitor in the marketing of biotechnologies . . . European nations have many fewer companies specializing in the marketing of biotechnologies. . . they do not generally dare to take industrial risks, therefore, they will not be as competitive as the United States and Japan.'

Why did such gaps in competitiveness and technology open in Europe during the seventies? Many reasons can be put forward, but all are connected with a common factor: European lack of cooperation. Competitiveness issues must be analysed chiefly at the level of the firm. Compared to their American or Japanese competitors, European firms suffered in the seventies from poor management and unfavourable environment, mainly due to:

Market size and heterogeneity
European firms are immersed in a fragmented marketing environment, and have generally been unable, during the past decade, to use the European dimension in activities where dimension is a key factor of success. When Japanese industrialists proved their ability to mass produce domestic equipment for the global market at low cost and good quality, European companies competed on their home markets, on a much smaller scale, and could not escape higher costs for no better quality.

Technophobia
In the days when Japan engaged in a massive technological build-up, Europe was cautious and reluctant about technology. This reluctance was apparent among the German, Dutch and Danish 'Greens' who opposed European nuclear power stations for environmental reasons as strongly as they resisted the occasional European breakthroughs in biotechnology for ethical purposes. Technophobia was also pervasive among European managers who had recently discovered that finance and marketing were better roads than R & D or manufacturing to reach the position of Chief Economic Officer (CEO).

Corporate isolationism
Most European companies failed to establish connections with two types of allies that were important actors in the success of American high-tech ventures because they brought in key ingredients for successful innovation: universities for fresh ideas and venture capitalists for fresh money.

Chauvinistic industrial policies

In this decade of unabashed governmental interference with business, most European nations had a tendency to promote their own 'national champion' for each industry.

This attitude was particularly dramatic for Europe's computer industry: in 1975 the French administration killed the Unidata project, formed two years before between CII (the French, and state-owned champion of the moment), Siemens, ICL, Olivetti and Philips in order to challenge IBM's leadership in Europe. The division of Europe between PAL and SECAM (respectively German and French standards for colour TV), a few years before, had proved favourable to the Japanese TV industry, but no one seemed to have drawn the lesson from this. Nor was the lesson remembered in the early eighties when German lobbies ruined the French attempt to establish a coalition between Thomson and Grundig that would have been Europe's last hope to produce its own VCR. Lack of cooperation between European markets, technologists, industry and universities, lack of cooperation between governments, Europarochialism and Eurorivalries have certainly been, in the seventies and early eighties, the main causes of Eurofailures.

One might object that in the same period European cooperation had been marked by significant progress, materialized for instance in the strength of Europe's common agricultural policy, the implementation of the European Monetary System, or the entry of Great Britain into the European Community. It is precisely because European cooperative efforts were concentrated on such priorities that technology and industrial competitiveness were very secondary issues. One has to bear in mind that Europe's common agricultural policy absorbs two-thirds of the EC's budget, while cooperative R & D programmes represent approximately 3 per cent of 'Brussels' resources! Moreover, since the Treaty of Rome and until 1985, the EC focused its R & D programmes on industries of the 'first industrial revolution' (coal and steel) with the notable exception of nuclear technologies.

Exceptions are indeed the best demonstration of what Europe needs in order to avoid technology and competitiveness gaps:

- with JET (Joint European Torus), Europe ranks first in the field of thermonuclear fusion (possibly next century's energy revolution);
- with ESA (European Space Agency), the European Ariane rocket had put in orbit in 1984 more telecommunication satellites than the American shuttle. (In 1989, Arianespace captured 50 per cent of the global market of satellite launches);
- with Airbus, Boeing began to face a serious challenger in the early eighties

All these examples prove that, even in the gloomy 1975–85 decade, European success could be achieved when two key ingredients were brought together:

technological cooperation among European firms or research institutes and favourable governmental posture.

Europhoria

Around 1985 a majority of Europeans became aware that major technological innovations among their triadic competitors frequently resulted from corporate cooperation and/or government encouragement. A revealing example is provided by one of Pilkington's directors during a conference organized by the European Industrial Research Management Association:

The USA and Japan have recognized better [than Europe] the leading role of technology in industrial competitiveness . . . they have better adapted their industrial structures to the modern technological age: in the USA through the Darwinian processes of the survival of the fittest and the birth of the new, and in Japan through that unique substitute for Darwin, MITI.[3]

It is certainly well founded to offer MITI as a model for industrial coordination. But other European analysts began in the early eighties to view the United States as no longer totally 'Darwinian'. Probably in reaction to the Soviet military threat, and the Japanese economic challenge, the officially 'liberal' American administration felt the need to manage the national endeavour of technological progress more efficiently. The Strategic Defense Initiative revealed to many Europeans the prominent role of the Department of Defense not only as a large purchaser of high-tech equipment but also as an orchestrator of cooperative efforts among American industry. The SDI was followed by cooperative projects in practically every domain in which the administration believed that the American technological position had to be strengthened: advanced computers (MCC consortium), advanced semiconductors (Sematech consortium), biotechnology (Human Genome Project) etc. The Reagan Administration went as far as to relax antitrust laws in order to allow joint-production ventures in the field of superconductors.

Encouraged by the Japanese and American example, Europeans took a radically new stance with regard to technological innovation around 1985. Having learnt from their own failures and competitors' exploits that cooperation was a key to technological success they took, with great enthusiasm, important moves in new directions.

Eureka

A direct, and non-military, answer from the French president to the American SDI, Eureka is a political initiative that unites nineteen European nations, and the EC Commission. The objective of Eureka is to raise, through closer cooperation among enterprises and research institutes in the field of advanced technologies, the productivity and competitiveness of Europe's industries on the world market. The main areas covered by Eureka are: information technology, robotics, new materials, biotechnology and telecommunications.

The principal characteristics of the initiative are the following:

- it is clearly market orientated, that is Eureka projects do not limit themselves to R & D cooperation; they encompass the full cycle of innovation, and must result in competitive new products (or services) addressed to the global market;
- it uses a bottom-up approach: 'to qualify as a Eureka project, companies from at least two European countries submit a detailed business plan to their Eureka national representatives. Once the project has been approved at that level, it goes before the plenary meeting of ministers from the nineteen countries who give the final go-ahead'[4]
- the financing of Eureka projects is the responsibility of the participating companies.

There is no such thing as a central Eureka funding authority. But when a project has received the Eureka label, financial assistance is more easily obtained from bankers and venture capitalists, and in many cases from individual governments. So far, public money has constituted approximately 40 per cent of the initiative financing. Eureka's impact is significant: by May 1989, over 800 companies and research centres have been cooperating in 209 cross-border R & D projects, worth more than $5 billion. Many other projects are under consideration today, proving that Eureka has undoubtedly gained a strong momentum throughout Europe.

The framework programme

A direct consequence of the European Single Act of 1986, the 'Framework Programme' represents the first concrete proof of R & D's legalization inside the EC's machinery. For the first time, Brussels' 'Eurocrats' have defined a clear and long-term ambition to bolster European R & D cooperation and industrial competitiveness through this programme. Within the five year (1987–91), $7 billion 'Framework Programme',[5] one can find 'Individual Programmes' relating to different technological areas:

- ESPRIT (European Strategic Programme of Research in Information Technology) is the largest, with nearly $2 billion for projects such as

VLSI (very large-scale integrated circuits), computer software, office automation;
- BRITE-EURAM is a $1,1 billion programme on new production technologies and new materials;
- RACE focuses on telecommunications with $600 million;
- thermonuclear fusion (JET-NET) disposes of nearly $2 billion;
- the rest is scattered among biotechnology ($330 million), health and environment ($550 million), programmes to stimulate scientific cooperation and researchers mobility ($340 million) . . .

The main characteristics of these programmes are the following:

- their goal is clearly not to transfer R & D projects to Brussels that are already under way around Europe, but to centralize any research work that can be more efficiently performed at the EC's level: for example programmes too costly for one country (nuclear fusion), programmes aiming at European standards or harmonization (telecommunications), programmes mobilizing synergies across national R & D projects (information technology);
- they must be 'precompetitive', which means that they must not lead to any cooperation in the industrial or marketing exploitation of R & D results, so as to avoid conflicts of interest between participating organizations, and not to infringe European antitrust laws;
- they must be transnational (involve at least two of the twelve countries forming the EC).
- they receive a 50 per cent funding for the EC.

By the end of 1988, 'Eurocooperation' was booming. Esprit's first phase had connected 3,000 researchers belonging to 450 different participating organizations (corporations, research centres, and universities) and now Brite coordinates about 300 trans-European projects. In the next 'Framework Programme', Esprit's funding is expected to double, BRITE's to triple.

Other Eurocooperations

Eureka and the 'Framework Programme' are not the only new cooperative efforts that have spread across Europe in the past five years, aiming more or less directly at the upgrading of European technology and competitiveness. (Not to mention the reinforcement of longstanding and successful cooperations such as CERN, Arianespace, Airbus etc.) These new projects belong to different categories.

Other EC programmes

Several programmes situated to promote and disseminate R & D have to be mentioned here, such as: COMETT (Community Programme in Education and Training for Technology) which aims at strengthening links between academic and corporate spheres, and accelerating the dissemination of new technologies across industry ($200 million for 1990–94) and SPRINT, the purpose of which is to facilitate trans-European technology transfers mainly among small and medium-size companies ($100 million).

Corporate Eurocooperations

All this heavily publicized EC and intergovernmental cooperative effervescence should not blind us to the intensification of technological cooperation between European firms, a more discrete phenomenon but probably even more meaningful for Europe's future. Brussels' programmes are estimated to represent approximately 2 per cent of total R & D expenditure in Europe. As an EC official recently declared: the Commission has only a catalytic role; we are not aiming at a massive subsidy policy. In our view, industry must itself remain responsible for the research it undertakes'.[3] Independently of Brussels, European firms have considerably increased their cooperative programmes in the past five years. Although it is impossible to provide precise quantification, it is estimated that 'independent' cooperative activities have largely exceeded in size those coordinated by the EC (thus demonstrating that Brussels' catalytic role has been efficient).

Though it is outside the scope of this chapter, the 'mergermania' that has swept Europe in the past two years is another proof of Europe's determination to recover (or defend) its competitiveness, and has had a strong impact on consequent technological cooperation. As recently explained in the American press: 'General Electric of Britain and Siemens of West Germany joined in a successful $3.1 billion hostile bid for Plessey—If successful, the takeover would create a European-based rival for America's GE and AT&T, and Japan's NEC and Fujitsu. More giant Eurocompetitors will soon be born. They will have both the will and the means to attack the Japanese and Americans in Europe, Japan and the US'.[6]

European companies are also encouraged to cooperate at a national level. If the days of the 'national champions' seem to have vanished, it is still the intention of many European governments to promote international economies of scale and encourage national synergies and cooperation. The French socialists have promoted the 'filière electronique', the German liberals the 'Fertigungstechnik' etc.

Last but not least, a good number of European companies today have realized the possible benefits of certain strategic alliances and have, of their own free will, increased their cooperation in technology (and marketing). Philips' international corporate research managing director commented precisely on that notion:

Cooperation to-day between industrial companies is not unnatural, it is more than fashion, it is more than a pleasant form of adultery or just a new wave or revolution, *it is a must.* Increasingly the required technology for one company cannot be supplied without cooperation with others. The technology is getting more refined, more complex, and changes in technology are faster than ever before. This calls for more expertise, more manpower in R&D, and in many cases, also for larger investments than companies can afford. All this can become obtainable through cooperation with partners.[3]

Small and medium-size companies (SMC)

Companies with less than 500 employees represent an important proportion of European industry (for example 50 per cent of West Germany's GNP and 66 per cent of its salaried workforce). Although they retain a minor share of European technological programmes, their role in Euro-cooperations is not negligible. SMCs participate in 22 per cent of RACE's projects, 25 per cent of BRITE's, 57 per cent of ESPRIT's and almost all SPRINT projects. Brussels has set up a 'task force' that encourages and helps cooperative effort, among European SMCs. Its most spectacular achievements are: the Euro Infocentres that provide all sorts of information and assistance to SMCs, including the field of technological cooperation and the Business Cooperation Network (BC NET), a computerized system that connects 340 advisers who dispatch thousands of requests and proposals for commercial or technical cooperation across the Community.

Company-initiated projects, nationally encouraged programmes, EC-subsidized programmes and other kinds of European initiatives really bring an optimistic note to European technological cooperation. Most of the former ingredients of Europessimism seem to disappear progressively: governments cooperate actively to achieve the great unified European market expected in 1992, a dialogue is undoubtedly being reinstated between business and universities and technology is much more fashionable among top management. Is Europe really going to take advantage of this flourishing technological cooperation?

Eurocontrasts

Europe has always been a field of complex associations, formed of complementarities and alliances but also of internal rivalries and antagonisms

'All that constitutes modern Europe divides it, all that divides it constitutes it'[7]. Edgar Morin's views about the permanent contrasts and divisions of Europe appear to be valid also when it comes to technology, cooperation and competitiveness. In this domain, the main European contrasts are the following.

'Fortress Europe' or 'open Europe'?

When Jacques Delors declared that he favoured a 'Europe ouverte, mais pas offerte' (an open Europe, but not one given away), he was probably trying to reconcile conflicting views about who should benefit primarily from Europe's accelerating integration: Europeans or non-Europeans? One might be tempted to take that dual approach when examining European technological cooperation. On the one hand, it is true that a good number of Eurocooperations, in the past few years, have been restricted to European companies, clearly joining in their efforts to compete better against non-Europeans.

A major example is the HDTV project (high-definition television). At stake is the replacement of 600 million TV sets world-wide. By 1986 Japan had nearly succeeded in imposing its norms on the entire planet. But Europeans agreed to postpone any decision on HDTV until 1990, and four major manufacturers (Philips, Bosch, Thomson, Thorn-EMI) plus twenty-five minor participants, joined up in a $200 million Eureka project to develop a competing European HDTV standard. The EC brought in its own contribution last November with a $50 million funding to promote and demonstrate to international potential users the superior quality of the European standard. Eureka's Secretary General declared: 'This one project justifies the whole Eureka initiative. In less than a year, the future of the industry has been rewritten.' Another significant example is the 'Mega project' of Philips and Siemens. Its purpose is to develop a submicron MOS technology for very large integrated, circuits, and improve the competitive position of European companies in this market.[3] The project's volume is approximately $1,5 billion ($750 million for R & D expenditures, and $750 million for investment in three factories). Philips and Siemens have obtained 40 per cent funding (from Dutch and German governments) of the R & D part.

On the other hand, facts and opinions also demonstrate that Europe is far from being a technological fortress:

- a good number of European companies have enjoyed triadic technological cooperation for many years with mutual benefits: Bull-Honeywell-NEC; General Electric-SNECMA; Philips–AT&T; Hoechst-Harvard.
- in the information technology industry, the number of technological agreements between European companies totalled 46 in 86, while those between European and American companies totalled 49.[8]

- European subsidiaries of companies such as IBM, AT&T and Hewlett Packard participate actively in several ESPRIT projects.
- prominent business leaders take a clear anti-Europrotectionnist stance: 'don't think only in terms of Europe . . .' (Hoechst, Dr Herzbrechstmeier); 'European collaboration in R & D must not preclude collaboration with Japanese and American firms (Pilkington, Sir R. Nicholson);[3] there are to-day more complementarity between European firms and US firms, than between European firms only' (Fiat, G. Agnelli).[9]

Our proposition to cut short the debate between any such things as 'Eurotechno-fortress' and 'Eurotechno-openness' is the following: if technological cooperation (strictly restricted to European companies) is capable of providing them with the best chances to compete on a global basis, then there should be no qualms about strict Eurocooperation (as in the HDTV case). This after all, is the prevailing attitude throughout the triad. If triadic R & D cooperation is likely to give better, quicker and less costly results than Eurocooperation, in that case there should not be any hesitation or Eurochauvism (the same cannot be said of non-European joint-ventures or investments—such as those undertaken in Europe by the Japanese car industry—that may be harmful to local manufacturers). That type of reconciliation in the 'fortress' debate is demonstrated by a number of European companies. Quite a few that have on one side, strict Eurocooperations, simultaneously carry on triadic technological cooperations on the other side: Philips competes aggressively with Japan in the Eureka HDTV project, but has also a joint-venture with Matsushita (electronic components and lighting devices) and cooperates with Sony on the compact disc standard. Siemens and Thomson (also participating in the HDTV project) have technology transfer agreements with Toshiba, etc.

Pre-competitivity or competitiveness?

It is required that projects in the EC's 'Framework Programme' are 'precompetitive'. This concept is ascribable to the will of most Eurocrats to increase cooperation between competitors while respecting the European antitrust regulations. Cooperation is therefore limited to research and must not be extended to product development, industrial exploitation or sales agreements. The invention of 'pre-competitivity' is fair but it also creates some controversy. The panel of industrialists and consultants in charge of the evaluation of the BRITE programme writes that they 'remain puzzled by the incorporation of pre-competitivity in a research programme whose primary goals are to enhance the competitiveness of European industry . . . Concern about the idea of pre-competitivity has also had the effect of pushing some projects upstream so that their industrial interest diminished.'[10]

Commission officials will object that they have often been generous in conceding exemptions to cooperative arrangements that go far beyond R & D, if company sizes and market shares met anti-trust criteria. 'Pre-competitivity' does not mean non-competitiveness. 'Framework Programme' contracts mention that if the research work is successful, the industrial and/or marketing exploitation of the results will have to take place within a 'reasonable' period, (that in no case can exceed ten years), and that otherwise the Commission may demand the reimbursement for the research funds it has granted.[11]

The real problem in our view is not the validity of the 'pre-competitivity' concept, it is in fact that a good proportion of European companies (40 per cent according to the BRITE report) have embarked on EC projects that did not form part of their overall research programme, knowing that 50 per cent of the R & D expenditures would be covered by EC subsidies. The real issue for European companies is better management of their technology, and improvement of coordination between the R & D, marketing and manufacturing functions. We totally agree with the recommendations of BRITE's experts that

industrial applicants should have to set up a structure for managing their technological resources . . . the consistency of the proposed project with the firm's major objectives will constitute a major selection criterion . . . to emphasize the importance of market orientation by industrial applicants, the Commission should require applications to be signed by top level managers and not only by R & D people.

Indeed 'pre-competitive' funding may be a Eurocratic issue, but competitiveness will always be a company issue.

Does European technological cooperation need to be supported by some kind of Euro-industrial policy or not?

A faction of European industrialists (and politicians) who agree on the reinforcement of European technological cooperation are opposed to anything that would resemble a Euro-industrial policy:

I doubt whether one can rely on Europe, let us say on the EC, to make an industry policy like countries do . . . I doubt whether the next step for Esprit has to be that we enter into big European industrial projects, technical integrated projects closer to the commercialization of products . . . this would mean that Brussels should decide which international consortia of companies have to be supported in their business and which not. That certainly is not a task for Brussels . . . Europe has to rely on the industrial companies themselves to make the join-ventures and coalitions they want', [Philips; Dr Kramer].[3]

Opposing that type of liberal approach, others object that when Europe becomes a unified market, it will be far more vulnerable to foreign penetration if it has not set up common rules and policies. Indeed until today many European industries have been state regulated, (telecommunications, space, arms etc.) and nationally protected. '1992' means the end of national state regulation, but not the end of national public procurement. In this type of industry, Europe as a whole needs some substitute regulation. Where can these new regulations and policies be designed, if not in Brussels where the Eurocrats have gained good experience in orchestrating the game between national governments and European industrialists? According to an adviser for it to the EC, the next phases of Esprit and Race should be much more oriented towards the definition of European standards and common strategies.[12] Bull's Chairman, J. Stern, also calls for a comprehensive European programme of standardization and public procurement in the field of computers.[13] Alain Minc explains in his last best-selling essay, that 'competitiveness and cooperation are inseparable; the Great European Market will demand accompanying policies'.[14]

Another article would not be sufficient to conclude the touchy issue of industrial policies, whether they are European, American or Japanese. But everyone can observe, in the domain of high-tech industries, a patent reinforcement of governmental coordination (and assistance) all across the triad. However, Europe does not possess any public institution with the economic and coercive powers of MITI or the DOD. Europe should try to catch up with Japan and the United States on that level, it would certainly take time . . .

With regard to technological cooperation and European competitiveness, the 'conclusion of the Brussels' official in charge of such matters sums it up well: both the Commission and industry are still struggling to improve management procedures. We are still moving up on a learning curve. But we can take consolation from Abraham Lincoln who, when his government was criticized for stumbling along, said "that may be true, but thank Heaven, we are stumbling in the right direction.'"[3]

References

1. M. Richonnier, *Les Métamorphoses de l'Europe*, Paris, Flammarion, 1985.
2. K. Ohmae, *Triad Power: The Coming Shape of Global Competition*, New York, The Free Press, 1985.
3. European Industrial Research Management Association Conference, Paper no. 34.

4. *International Management,* Eureka reprint.
5. Recherché et développement technologique pour l'Europe, *Dossiers de l'Europe,* 1987.
6. *Fortune,* 19 December 1989.
7. Edgar Morin, *Penser l'Europe,* Paris, Gallimard, 1987.
8. ESPRIT, *Annual Report,* 1987.
9. *Le Monde,* 28 March 1987.
10. BRITE. Evaluation report, 1989.
11. J.M. de Leersnyder and Jthieffry, *L'enterprise face à l'Europe,* Paris, Dunod.
12. A. Danzin, *Common Market Review,* no 307, May–June 1987.
13. *Le Monde,* 30 March 1989.
14. Alain Minc, *La grande illusion,* Paris, Grasset, 1989.

7 Defence technology and European security in the 1990s: A British perspective

Christopher Coker

Admiral Gorshkov is alleged to have said that the side that wins the next war will be the one that makes the better use of the electro-magnetic spectrum. Given the Warsaw Pact's present superiority over NATO in men, platforms and weapons, it is essential that the Alliance maintains its *qualitative* edge through the use of advanced technologies. Unfortunately, for the first time in the Alliance's history, Europe is now fighting on two fronts, one against the Soviet Union, a battle which it is likely to win, the other against the United States, a battle which it is already losing. At a time of proposed quantitative force cuts, quality is likely to count more than ever.

The most rapid advances are being made in information technology (IT), not only in integrated circuitry but also in intelligence, knowledge-based systems and the man machine interface. Unfortunately, in IT the Europeans find themselves in a desperately weak position. At present they only have 5 per cent of the world market in integrated circuits. In Europe, production of the 265K memory chip, which has been in production in the United States and Japan since 1982, has only just begun.

The position is little better in other areas of technology where crucial advances are already occurring. 'Stealth' technology has succeeded in reducing radar cross-sections (RCS) to the point where the B1-Bomber is alleged to have an RCS only half that of a Cessna 172, the aircraft piloted by Matthias Rust in 1987 in his flight to Moscow. Whether or not this is true, the B2-Bomber has a very small RCS indeed. Yet in Stealth technology, Europe is nowhere to be seen. One of the reasons Britain chose to develop the European Fighter Aircraft (EFA) at an absurdly expensive price was the refusal by the United States to sell the F-117 for fear that its Stealth technology might end up in Moscow. Doubtless the claims that a Stealth fighter would lack the agility and air configuration of a conventional plane influenced the decision. It influenced it, however, by default, given that Stealth technology was not only beyond Europe's grasp, but also in the gift of a country that was not prepared to provide it.

It is clear that the United States is increasingly involved in a series of technologies which are beyond Europe's technical competence. This is particularly important in a third area of activity: micro-electronic components, which are vital for the production of super-computers and other defence equipment. Even the United States is having difficulty competing with Japan. In the latter case, from figures provided by the United States Board of Army Science and Technology, it has been estimated that 40 per cent of all advanced electronics in American weapon systems are Japanese made. In a move prompted by increasing security concerns about over reliance on Japanese manufactures, the United States National Security Agency (NSA) has decided to produce its own integrated circuits, beginning with a $85 million contract to build a six-inch (153mm) silicon wafer plant in Fort Meade, Maryland.

In these three critical areas—Stealth, micro-electronics and information technology—Europe's weakness is unlikely to be offset by a less-fragmented market after 1992. In an attempt to protect the European market, the Eureka programme has targetted such specific areas as fifth-generation computers, 64-megabit silicon-based memory chips and 'intelligent' sensors as the key technologies which will enable the Europeans to break through into the third industrial revolution—without which they will not break out of the second. British doubts about Eureka are that the initiative is a form of protectionism which is unlikely to impress the United States. At a meeting of the European Council in September 1983, the French put forward a proposal that European high-tech industries needed to be protected for at least five years if only to enable them to reach international standards of competitiveness. In the security field the situation has not improved for the better. While the United States and Canada account only for 1.5 per cent of world weapons imports, their West European allies import five times that value in defence equipment, most of it American in origin.

Whether Britain is right in its perception that Europe should not compete with the United States, the United States is aghast that the Independent European Programme Group (IEPG) has begun to take steps towards the creation of a European armaments market. In particular, an action plan approved by defence ministers in November 1988 recommended measures to enable more 'border-crossing competition' for contracts (including the publication of lists of contract opportunities) and steps towards more cooperation over research and technology. Once a half-comatose organization that met at the ministerial level only for the last four of its twelve years and then irregularly, the group is now due to get a permanent secretariat to organize greater 'cross-purchase' buying—buying in each other's markets on a reciprocal basis as the French and British have recently been doing.

In March 1989, Mack Mattingly, Nato assistant secretary general for defence support, suggested that if the IEPG continued to exclude observers from North America, it could accentuate the growing perception that the IEPG had become an 'inward-looking, cosy Europe-only club, not an outward-looking force and contributor to wider Atlantic cohesion'. These allied differences are unlikely to be resolved soon. In fact, a joint European defence industry study drafted by representatives of six European firms, while welcoming the plan to open up competitive bidding for defence contracts as a useful exercise, claimed that it fell far short of a solution to Europe's problems. The report wanted the IEPG's machinery to be 'radically overhauled and upgraded' and its role to be 'drastically strengthened'.

It is not clear why. As the IEPG Vredeling Report *Towards a Stronger Europe* pointed out, most companies are opposed to developing further production capacity in Europe's defence industry. Indeed, there is already excess capacity in the weapons sector, now that exports to Third World countries have begun to decline.

The real problem is that Europe's vulnerability may increase still further, that Europrotectionism is not the answer, that protectionism on the part of the United States, which is no less real for being undeclared, may well decide the issue in America's favour.

1992: American Challenge to Europe

America's challenge to Europe takes many forms. American expenditure on R & D (both private and public) far outstrips Europe's, 83 per cent of which is accounted for by Britain and France. Over the next few years, the American government expects to spend about $200 million on superconductivity as well as $2 billion this year on twenty-one of the twenty-two key technologies it considers critical to the long-term superiority of American weapons systems. Although the budget for some technologies—such as those involved in suppressing the radar signature of weapons systems—remains classified, others such as the development of sensitive radars and computer modeling are funded out of the Strategic Defense Initiative (SDI) budget. Congress has already intimated it may move some of the programmes out of the latter in order to protect them from SDI cut-backs.

In addition, American technology is no longer available on such easy terms. The day when countries such as Japan could develop a semi-conductor industry with American technology and then break into America's own market with such devastating effect has passed, perhaps for ever. When the (American) Motorola Electronics Company announced a new generation of microprocessors in 1984, it added that it would be selling no more licences and that it was interested solely in technology exchange. The problem is that Europe has little technology of its own to exchange, at least in a growing number of areas specific to its own security.

It can be claimed that the Balance Technology Initiative (BTI) established in May 1987, a programme which concentrates on advanced conventional force multiplier technology, involves a significant American contribution. The United

States has promised to spend more than $1.5 billion on joint American-European BTI projects by 1993. It can also be said that NATO's Conventional Armaments Planning System (CAPS) plans to foster joint American-European co-production to exploit 'collective industrial strengths' and to provide a framework for developing armaments' plans consistent with NATO's long-term planning, in other words to bring NATO's tactical requirements and its defence industry capabilities more sharply into line. CAPS, however, has only a two-year life span. To make matters worse, the changes may be well-meaning but they are structural and possibly cosmetic. Their very acronyms, BTI and CAPS, suggest the modernization of an existing system, little more than plastic surgery, an attempt to provide a new face for something that has existed all along. Most of the projects listed under the BTI scheme were funded well before the new umbrella acronyms were created.

Third, like other foreigners, the British have found themselves increasingly discriminated against even after buying American companies. American firms passing into British ownership have frequently been debarred from further work on projects that have already been initiated. British businessman have frequently found themselves banned from conferences on sensor technology. When Britain and America *have* teamed up, British companies have been denied adequate access to America technology, even when companies in the United Kingdom have been sent components for assembly prior to transmission back to the United States.

Nothing has concentrated Europe's mind on this problem more than the SDI project. Where until recently Britain and its partners parted company was the belief on the part of the British government that the SDI programme might mark a breakthrough. It was for that reason that it was the first European government to sign a Memorandum of Understanding with the United States in 1985 allowing British companies to tender for projects in eighteen different areas of SDI research. Of all countries, Britain should have been in the best position not only to take the lead but also retain it. It already enjoyed the benefits of another, and much older Memorandum of Understanding which placed British manufacturers on an equal footing with their American counterparts when bidding for Defense Department contracts. Over the years, the special relationship (such as it was) had also produced regular returns in the scientific field, particularly on charged particle beam research at Malvern.

Mrs Thatcher's government was convinced that if the SDI programme was to be realized successfully the United States would need British technology, especially in the fields in which the United Kingdom was still the world leader, such as ultra-high speed optical computing. The question now remains, was the United Kingdom correct in its assessment?

The answer would appear to be no. The British did not help by failing to fund areas in which they did have a lead over the United States, particularly British Aerospace's Hotol programme, a horizontal take-off and landing shuttle based on a revolutionary new propulsion system which might have given the United Kingdom a chance to verify arms control agreements without reliance on the United States. The British government's refusal to fund the Hotol project was probably one of the worst examples of short-sightedness in recent British history.

The SDI programme has shown that while the United States *has* made use of British technology—optical computing at Herriot Watt and neutron particle beam technology at Culham—the British themselves have simply become sub-contractors to the United States in most other areas. Far from realizing high returns, the United Kingdom has received less from its SDI contracts than Israel. It has become clear that rather than give the Europeans project leadership in any part of the SDI programme, the United States is only prepared to parcel out individual research commissions which have had very little long-term impact on the standing of European high technology. Indeed, Senator Glenn's legislation now makes it practically impossible for the United Kingdom or any other European country to have the access to SDI technology which was implicit in the agreement of 1985. As usual, Congress is likely to remain considerably more suspicious than the administration about transferring sensitive technology to potential European competitors.

It is, of course, a development which invalidates the case for greater burden sharing within the Alliance. Burden sharing means power sharing, and power sharing these days means European access to the American defence industry, not merely investment in those pharmaceutical or fashion companies which have now made the United Kingdom America's largest foreign investor, well ahead of Japan.

Fourth, European companies cannot expect to match the economies of scale found in the United States and Japan, something which 1992 is not going to change significantly. In particular, the cost of developing new products is now well beyond the financial resources of any one European concern. The development costs for the next generation of main switching-in telecommunications (an essential element of the battlefield communication systems) could be $1 billion per product. In order to compete, or simply survive, Europe's nine leading companies will have to work together. EC-1992 does not suggest the process will lead to such cooperation, or make it inevitable. By comparison, the American market is served by only three corporations, all of which are able to absorb high development costs by virtue of their size.

It is a point well illustrated by the battlefield communication system RITA (French-designed) for which the United States Army opted instead of the British system Ptarmigan for its Mobile Subscriber Equipment (MSE). Although hailed at the time as a major European contribution to American defence, it has not proved a new model of transatlantic cooperation. Instead, the American partner GTE gained 70 per cent of the MSE contract, a higher percentage than Rockwell could have expected had it cooperated with Plessey on Ptarmigan. In the end, this was one of the central considerations which persuaded the Americans to choose the French system.

Fifth, will European companies survive the large volume of mergers and acquisitions which are proving an important aspect to the run-up to 1992? At present there is not a single private organization in Europe that can speak with a common European voice on defence technology. The only vital common programme available to European firms is the United States/non-United States MOU Cooperative Defence Acquisition Process. A question must be asked about the adequacy of existing regulatory regimes for controlling mergers and maintaining competition. At the national level, bodies such as the Monopolies and Mergers Commission in the United Kingdom, are empowered to make recommendations to ministers before a merger has been completed. There is no equivalent body able to take a Europe wide view. The European Commission has only a very limited, and still evolving, power to investigate mergers and collaborations. It is excluded under the Treaty of Rome from involvement in matters of national security. The Single European Act (1987) contains no explicit revision of this position, although it does state that the contracting parties are determined 'to maintain the technological and industrial basis necessary for their security'. Although the Commission therefore does have some influence in the defence field, it does not have enough to regulate competition.

A clear example of this is telecommunications. With the relative demise of its computer and semi-conductor industries, telecommunications remains one of the few viable sections in European emerging technology. In 1985 the British government was only prevailed upon by the European Commission to cancel a contract between British Telecom and IBM to operate a data communications network in the United Kingdom when it was pointed out that had the deal gone ahead, it would have been an important step in IBM's stated intention to dominate completely the European market.

Finally, 1992 will probably be bedevilled by the old problem of national security. European cooperation may be necessary in high-tech defence industry, but this does not suggest that European *states* will necessarily become more open to arguments in favour of defence interdependence. At a meeting in November 1985, the British minister in charge of technology, Geoffrey Pattie, talked of the Eureka programme providing 'a sharp focus' for European collaborative efforts. It is significant, perhaps, that the British have described Eureka as a 'searchlight', identifying promising areas of collaboration, not, in the words of the French, as 'an accelerator' between the drawing-board and the market-place.

It is a difference which illustrates that the British and French are still at odds about their attitude to private financing. The present British government, true to its economic first principles, clearly views Eureka as a pragmatic exercise in collaboration by private industry to which governments can best contribute by dismantling internal trade barriers rather than providing subsidies. Inevitably, this market-orientated approach has been taken as further evidence of lack of enthusiasm for yet another European programme, a typical holding operation by a country whose commitment to European procurement can be measured by its marginal investment in the European space programme—if its investment can be called even that.

It is an attitude, however, which is probably overly cynical. The differences between Paris and London are genuine disagreements of principle. The British wish to tackle the problems from the demand end, to identify the sectors where coordinated research could pay off exceptionally well. These philosophical differences between demand-side and supply-side economics are unlikely to disappear after 1992. In the security field, they will be increasingly highlighted as Europe finds the task of competing with the United States increasingly difficult.

Does Europe have a role?

The only area in which real progress in European defence cooperation has been made is that of defence procurement. The creation in 1976 of the Independent European Programme Group (IEPG), which was formally outside NATO, offered a way forward for the Europeans even if it did not live up to the promise of the Klepsch report (1978) which would have made the IEPG accountable to the European Commission and established, in the process, a distinct European presence within the Alliance rooted in an organization outside NATO.

David Greenwood hoped at the time that the IEPG would one day be able to define a 'distinctly European' theory of defence and provide the equipment for it. He recognized that although this might fall short of a wholly independent European defence organization it might make Europe considerably less dependent on Washington. Whether this is any longer possible is a moot point. The new conventional doctrines such as Follow on Forces (FOFA) which have come into focus in recent years would make Europe much more dependent on American technology. Few of the deep-strike and FOFA technologies are European. It is difficult to imagine on tomorrow's battlefield a 'distinctly European' theory of defence, only American theories applied to European conditions. Perhaps this is further evidence that a specific European defence identity may no longer have a future, only a past.

Having drawn this pessimistic view, it is only fair to point out that the situation for Europe is not entirely depressing. By nature, the Europeans do not tend to think in terms of technology when confronting military problems. Indeed many of the technologies in which the United States places its hopes for the future are well 'over the horizon' not 'on the horizon' as the American strategic community often claims. Even the technologies which Europe possesses such as Marconi's process in thermal imaging, the production of the multi-rocket launch system

MRLS-3 and the milimetric radar seeker by GEC are unlikely to be integrated into *force planning* until well into the 1990s.

Recent procurement examples in the United States suggest that America may be a long way off from deploying the new technologies. The American JSTARS (Joint Surveillance Target Attack Radar System) has encountered major problems. As is often the case, software faults in design seem to be at the heart of the matter. The Alvey Report on IT in 1985 quoted data for federal software projects which showed that only 2 per cent of the software was used as delivered; 3 per cent was used after minor modification; 19 per cent was abandoned or re-worked; 29 per cent was not delivered and 47 per cent was delivered but not used.

In a word, the unreliability of American technological breakthroughs in the security field clearly suggests that Europe will not be dominated by the United States any more than the Europeans can hope to remain independent of the United States. Back in 1977, Geoffrey Pattie confidently argued that the new technologies would provide 'a new role for the medium power'—that new electronic devices and precision-guided munitions might present the wherewithal for a European 'challenge' to both super-powers. 'Translated into military and diplomatic terms', he argued, 'we are entering a phase where inadequately prepared powers of Britain's size can recover some of the diplomatic leverage that massive nuclear deterrence destroyed in the 1960's.'

It was an alluring but short-lived vision. Now that Europe's technical shortcomings are becoming all too apparent, it is a vision which has not only faded but vanished altogether. While Europe may be able to develop, for example, one of FOFA's requirements, the Strike Shallow and Selectively FOFA System, it is not technologically able to produce a system capable of operating at a greater range.

On the American side, however, it is clear that the technologies that might make conventional deterrence a reality are also many years off even if some of the systems are in production. The Alvey Report suggests that the United States would do well to remember that the premature introduction of the tank in the First World War, like the German V2 rocket in the Second, illustrated how the eternal quest for technological dominance can provide almost unlimited scope for ineffectiveness.

Part IV The Tripartite Relationship

8 European management of trilateral interdependencies

Gavin Boyd

The European Community is evolving into a multicountry political economy with confederal institutions for collective management. From an intermediate stage of economic integration, at which large trade, transnational production and financial links have developed between the member states, advances are being made toward economic union. The formation of a common market is being completed, and a strengthening of the European Monetary System is preparing the way for the establishment of a European central bank and a common currency. Community decision processes are being strengthened to facilitate more active cooperation between the member governments and extension of their cooperation into areas where it will be needed to attain all the expected benefits from full market integration. These areas include industrial and technology policy and may be extended to include foreign direct investment policy.

The most important process in the advancement to a higher stage of economic integration is the current removal of non-tariff barriers to commerce within the Community, to be completed by 1992. This endeavour, authorized by common adoption of the Single European Act in 1987, has been a response to the growth retarding effects of forms of market separation persisting in the Community and to losses of competitiveness by Community firms to American and Japanese enterprises, in external markets as well as within the Community. The problem of European international competitiveness had assumed great significance because of the depreciation of the US dollar which began in 1985 when the United States switched from a *laissez-faire* to an interventionist exchange rate policy and secured cooperation for this from the other members of the Group of Five. The depreciation was intended to serve trade policy objectives, and trade policy meanwhile was being affected by strong protectionist pressures in the Congress, directed in part against the Community. At a more fundamental level, moreover, imperatives for more self-reliant as well as increased growth in the Community were becoming evident because of the United States' serious macromanagement problems, caused mainly by heavy government debt which were dramatized in late 1987 by the American stock market collapse. These problems posed great uncertainties about the future growth effects of Atlantic commerce and cross investment, and indicated that the Community would have

to strengthen its capacities to adapt to shocks in what was tending to become a strained relationship with the United States. The complete integration of the internal market, it was clear, would aid the growth of such adaptive capacities, not only by strengthening the economies of the member countries and linking them more closely, but also by giving impetus to processes of economic cooperation among the member governments.

Questions of trilateral economic cooperation have been posed with increasing urgency while the drive for complete internal market integration has been under way. The Community has had to contend not only with losses of competitiveness in global markets and American protectionist pressures, but also with issues concerning the United States' role in trilateral relations and its related involvement in the international political economy. Within the trilateral context the United States relates very actively, from a position of considerable strength, to both the Community and Japan, but mostly in separate dealings, while the Community and Japan relate to each other rather distantly. Forceful use of American bargaining power is thus possible in the separate interactions, and in the Atlantic setting this is facilitated by insufficient progress toward institutional integration within the Community. Differences between West Germany, France and Britain, the three major members of the Community with whom the United States interacts as individual states in the Group of Five, tend to prevent full utilization of the Community's potential for collective bargaining in Atlantic relations.

In the international political economy America's long-standing commitment has been the building of a liberal trading system, but this has been altered by the strong protectionist pressures influencing her trade policy, and by a new concern with realizing trade surpluses needed to help service foreign debts. These changes, together with the interventionist foreign exchange policy, are threatening to have destabilizing consequences. More serious destabilizing consequences are being threatened by strains in the American economy resulting from heavy debt burdens and decreases of business confidence. Further depreciation of the dollar is possible and could be accompanied by a recession.

The protection of European interests while greater international competitiveness is attained through complete internal market integration requires more active European engagement with the tasks of managing trilateral interdependencies. A stronger and more constructive European role is needed to promote more genuinely collective trilateral management and to ensure that such management will be functional in terms of the Community's needs for more balanced and less asymmetrical interdependencies with the United States and Japan. The structural interdependencies are increasing with greater asymmetries because of the competitive advantages of American and Japanese firms in trilateral commerce and transnational production. The interdependencies in policymaking are also increasing asymmetrically because

Community decision processes are being conditioned more by actual or prospective American and Japanese policy choices.

The Community's capacities to manage its structural and policy interdependencies with the United States and Japan are affected by problems of internal management, deriving from a lack of integrative behaviour in the mutual relations of its members, especially the larger ones in the Group of Five, which tends to slow the pace of advancement toward economic union. At present the area of internal collective management is not being expanded with the speed and efficiency needed to ensure that the benefits of full market integration will be realized sufficiently by European firms, instead of being captured by American and Japanese enterprises exploiting the scope for entries and expansion allowed by the presently uncoordinated foreign direct investment policies of the member–states.

The Community is also disadvantaged because, while the drive for complete market integration is contributing to increases in political cohesion between its members, these are not sufficient to overcome the differences between those members over external issues which give bargaining advantages to the United States. These are differences in policy orientations resulting from the contrasts in distinctly national concerns of policy communities in those members, influenced by cultural factors that are resistant to change. The development of Community-wide societal integration, which will tend to be aided by the complete merging of national markets, will over the long term assist the growth of political cohesion between the member countries, but this could be set back by conflicts over the spread of gains from market integration. That risk, and the need for advances in political integration to ensure more effective management of relations with the United States, are making the emergence of strong leadership within the Community more and more necessary for its evolution as a structure for collective European management and collective international management.

The Community in the world political economy

The European Community is a confederal grouping of culturally affinitive states at relatively advanced levels of political evolution and intermediate and advanced levels of industrialization. The degrees of political development and the prospects for further political evolution are affected by varying problems of governance. These are manageable and do not significantly affect political continuity in West Germany's neocorporatist system, which sustains a highly efficient form of organized capitalism that dominates the Community.[1] The problems of governance in Britain's pluralistic system however are serious, especially because of the persistence of quite divisive social polarization, and Britain's economy, which has experienced considerable deindustrialization, continues to lag behind West Germany's.[2] France's pluralistic system is somewhat less affected by problems of governance because of degrees of policy consensus and party convergence resulting from cooperation between a Socialist president and a conservative cabinet during the mid-1980s. The French economy has been burdened by a large public sector since the initial period of Socialist government during the first half of this decade, but industrially is lagging less than Britain behind West Germany.[3]

The political and economic attributes of the three large West European industrialized democracies, and their policies, are major determinants of the Community's evolution. The strong, stable and highly advanced West German political economy is a dynamic core area of the grouping, and it provides leadership, in cooperation with France, for deepening economic integration. Franco–German trade is the largest volume of commerce within the Community, and for Britain, Italy, the Netherlands and other Community members, West Germany is the main European trading partner. The central German involvement in Community trade, linked with very extensive activity in world trade, sustains a dominant role in the European Monetary System, under which other European currencies are in effect linked with the German mark.[4] In the current drive for complete market integration France operates in partnership with West Germany while ensuring the cooperation of Italy and the smaller members. Britain, because of long-standing reluctance to become closely associated with the movement for West European integration, remains somewhat apart from the Franco–German leadership coalition, and, while not participating in the European Monetary System, is a source of opposition to the establishment of a Community central bank, which the West German and French administrations wish to establish on the basis of cooperation received from other members of the European Monetary System.

The Community endeavour to achieve full internal market integration has been the outcome of slow collective policy learning and slow collective decision-making. The economies of most of the members began to stagnate after the first oil crisis in 1973, when the negative growth effects of high energy prices were increased by national resorts to nontariff barriers which raised levels of market separation within the Community. Adjustment to these problems was not feasible for most European firms, and they were further disadvantaged by the second oil crisis and the sharp rises in American interest rates toward the end of the 1970s. Industrial restructuring of the kind being achieved at that time by Japan was difficult for most of the Community members and to a degree even for West Germany. Losses of competitiveness in external markets followed, while the Community market was further penetrated by American and Japanese enterprises. The Community decision structure did not respond until 1985, when the European Council accepted a plan for the elimination of internal trade barriers by 1992, and implementation of this did not begin until the general adoption of the Single European Act in 1987.[5]

Broad consensus on the need for complete market integration has developed among national policy makers, and there has been an advance in institutional development which allows majority decision-making in areas directly related to the removal of internal trade barriers, but a common political will to undertake wider-ranging cooperation for management of the integrated market has not yet developed. The European Parliament has been given a qualified role in the Community's decision processes under the Single European Act, but lacks authority to give an integrative thrust to the activities of the Council of Ministers. Members of that body act primarily as representatives of their national governments, and these evidence much concern with retaining autonomy in technology, industrial, foreign direct investment and external trade areas so as to have some control over the spread of gains from fully liberalized intra-Community trade. The potential common benefits of more comprehensive collective management are clearly not expected to be attainable because of perceptions of the problems of establishing an adequate collective decision-making system, and scope for some defensive and promotional measures that can affect the distribution of benefits from market integration is understandably seen to be desirable.[6]

Neofunctional logic is thus tending to be frustrated, and this problem may well become more serious if strains in the Community's external economic relations reduce the growth effects of internal market integration and increase asymmetries in the spread of those effects. Those asymmetries, resulting from intensified competition among Community firms, and between Community and outside firms, for shares of the integrating market, are likely to be substantial because of the failures of Community members to move toward more comprehensive collective management. The involvement of large American international firms is likely to be especially active because the extensive American corporate presence in Western Europe is well positioned for further expansion. The main concentration is in Britain, but this is not growing as fast as American direct investment in West Germany.[7] British reluctance to cooperate with Community partners for deepening integration provides a degree of insurance against possible attempts by those partners to introduce common industrial and foreign direct investment policies that could limit opportunities for outside firms. For the present many Community administrations are competing against each other to attract direct investment from their partners in the grouping and from outside states by offering tax concessions and subsidies.

In the international political economy the Community's involvement is based more on arms-length trade than overseas production because of social and political restraints on outward direct investment and because such investment is discouraged in the various regions where market shares have been lost to American and Japanese firms.[8] Community firms have incentives to move operations abroad because of the costs associated with welfare-state burdens in Europe and the degrees of economic stagnation in the Community, but resources to support outward direct investment are lacking on account of that stagnation and the continuing effects of the forms of market separation within the grouping that are being slowly removed. Competition in global markets is experienced primarily from the exporting and transnational production activities of American enterprises. American exporting has become highly competitive because of the depreciation of the dollar, while the larger American involvement in foreign markets through overseas production has continued to grow, despite the currency depreciation, because of the magnitude of resources already available from overseas operations and because of the prospect of stresses and higher government costs in the United States until its large budget deficits are overcome. Meanwhile Japanese exporting, despite the appreciation of the yen, has remained intensely competitive because of rapid industrial restructuring at home and a superior capacity to accept lower profits in order to expand market shares. American, Japanese, and East-Asian NIC penetration of the Community market, moreover, has increasingly limited the utilization of its potential by Community firms and has thus affected their capacities to operate in outside markets. Weaker and less innovative firms in the Community have been forced into cumulative declines by the external competition and by the operations of more dynamic Community firms moving into their environments.[9]

   In the international trading system the Community functions as a loosely coordinated group because of the efforts of the larger members to maintain high degrees of autonomy in managing their commerce with outside states while cooperating with each other in support of the grouping's common external tariff and of its informal external trade measures, including 'voluntary' export restraints imposed on Japan.[10] The member–states manage their own trade policies, with the use of formal and informal measures, while participating individually in the activities of the General Agreement on Tariffs and Trade (GATT). Community interest in the GATT system is not very active because it tends to be seen as a bargaining forum in which the United States has major advantages on account of its bargaining power and the difficulties of evolving common Community positions on external trade issues. The United States, moreover, is recognized to be under strong domestic pressure to secure wider access to the Community market in order to reduce its trade deficits. To develop a stronger role in the international trading system the Community uses preferential arrangements with groups of trading partners, but these have been of limited utility in the overall competition against the United States and Japan. The most ambitious scheme has been the special relationship with the African, Caribbean and Pacific States, implemented under successive Lomé agreements, and it has encountered difficulties because of stagnation in many of the African countries.

   In the international monetary system the Community has no formal role to complement its official responsibility for managing an external commercial policy. West Germany and France relate to the United States individually and as members of the European Monetary System, while Britain interacts with the United States exclusively on an individual basis but with a negative attitude to the European Monetary System that hinders its development and prevents collective European management of monetary issues in Atlantic relations. West Germany dominates the European Monetary System, managing its own monetary policy independently, with a restrictive emphasis that restrains inflation and internal demand, while France and the other members of the system converge toward the German policy.[11] A considerable degree of shared protection against dollar fluctuations is thus possible, and the system provides a zone of stability which is especially, important for the large volume of intraCommunity trade. United States pressure is applied against West Germany to induce shifts to an expansionary policy, but the strong West German position within the Community facilitates resistance against this leverage, despite indications that West German fiscal and monetary restraint imparts a degree of deflationary bias to the Community.

The Community's growth prospects depend in a large measure on improved performance in the international political economy. This, however, is likely to remain a loosely coordinated endeavour rather than an increasingly collective effort, with West Germany strengthening its leadership in the European share of world commerce while France and Britain fall further behind because of slow technological and structural adjustment. Heavy dependence on exports to the United States makes each of these states and the Community as a whole vulnerable to fluctuations of the dollar in currency markets and to changes in American trade policy, as well as to any slackening of growth in the United States.[12] Community interest in commerce with the Soviet Union and the East European states is being encouraged by changes in those regimes, but this trade has been on a small scale for many years and is not likely to increase significantly unless there is substantial economic liberalization in the Soviet Union.

Atlantic interdependencies

Large structural interdependencies have evolved between the European Community and the United States, mainly because of the expansion of the American corporate presence during the Community's initial period of rapid growth before the first oil crisis, but also because of increases in Atlantic trade which were associated with that growth. Slackened growth in the Community has tended to slow Atlantic commerce and the expansion of American firms in Europe, but the incentives for such firms to extend their operations have been increased by the current drive for complete market integration. Meanwhile, influenced apparently by the slower growth in the Community over the past decade and a half, European firms have been increasing direct investment in the United States, subject to the social and political constraints which tend to bind them more closely than American firms to their home economies. The appreciation of the dollar during the first half of this decade encouraged this cross-investment. [13]

The structural interdependencies have been firstly trade and production linkages, through which firms from each side of the Atlantic relationship have become increasingly involved in the operation of the other side's markets, in competition against and also in collaboration with the other side's enterprises. Large asymmetries with cumulative effects have developed because very extensive shares of Community markets have been gained by American firms, especially those producing in the Community, while shares of the American market gained by European firms have been relatively smaller. The contrasts have had increasing effects on the overall growth of the two types of firms, causing greater and greater disparities in size, resources, technological levels, productivity and capacities for expansion. The American firms, gaining advantages from their more extensive European and global operations, and from higher levels of spending on research and development, as well as from greater financial resources, have been able to combine rationalization of their Community activities with the evolution of their overall international strategies. In addition these larger outside enterprises have been able to cope more effectively with the slackening of growth in the Community, which has hampered development by firms based in the member countries, and have been better placed to reduce exposure to host country taxation. [14] Within the Community the main competitors have been West German enterprises, which have relied heavily on exporting to other member states from their home base and have had less scope to move production processes into those other members for nationally responsive strategies, partly because of political factors. Competition from French firms has been on a much smaller scale, although political problems appear to have limited direct American investment in France. With the decline of manufacturing in Britain competition from that country's enterprises has been of minor significance, and indeed the decline appears to have facilitated expansion of the American corporate presence in Britain larger than that in West Germany but growing at slower pace.

Community firms moving into operations in the United States have mostly been disadvantaged by the limitations of their smaller home country markets and have had to contend with the strong oligopolistic positions of many host country firms in the integrated American market. [15] Substantial flows of Community direct investment into the United States did not develop until the 1960s, and although these have grown to levels which now match outflows to the Community from the United States their impact on the American economy has been relatively small because of its size. The growth of the Community corporate presence in the United States moreover, is on a smaller scale than the expansion of the American multinational presence in the Community because this is growing much more through the reinvestment of profits. Political risk factors in the home environment have motivated some of the Community direct investment in the United States and may well continue to be significant because of the establishment of a second Socialist administration in France and the persistence of problems hindering industrial recovery in Britain. In recent years the depreciation of the dollar has facilitated increases in the flow of direct investment from the Community. The largest increases have been in the volume coming from Britain, and British firms are now well established as the main foreign manufacturing presence in the United States, being more than twice the size of the total West German and French involvements in manufacturing there and more than four times larger than the Japanese manufacturing presence.[16] Earnings by Community firms from direct investment in the United States appear to be about one-fifth of the earnings of American firms from direct investment in the Community.

The asymmetries associated with Atlantic cross-investment have resulted from the mostly independent activities of firms, especially those operating out of the United States and from the influence of America and Community policy mixes on those activities and on the operation of markets. Decisive factors have been the early large-scale entry of American enterprises into the Community, during its initial growth period, and the subsequent increases in nontariff barriers separating, in varying degrees, the markets of the Community members. The development of Community firms has been seriously hindered by those barriers, but they have been less significant obstacles for the larger American enterprises, which have been able to utilize more extensive resources and spread their operations on a wider scale, while benefiting from the inducements offered by European host governments and from the interest of host country firms in joint ventures and linkages.

Large-scale arms-length and intrafirm trade has been associated with the transnational production linkages, adding to the dimensions of Atlantic structural interdependencies. The intrafirm commerce has been estimated to be about 50 per cent of the United States' exports, and, as it involves transfer pricing, results in undervaluation of those exports. Intrafirm commerce appears to represent a lower proportion of Community exports to the United States, and this must result in a similar undervaluation, but on a smaller scale. Until the dollar appreciation during the first Reagan administration, the United States usually had a substantial surplus in Atlantic trade which was not fully reflected in official statistics because of the volume of items covered by transfer pricing.

The development of Atlantic trade has been influenced more directly and more substantially by administrative measures, on each side, than the evolution of transnational production linkages, which have been shaped mainly by the independent decisions of firms operating within the framework of neutral foreign direct investment policies. The Community's high level of farm subsidization and protection has severely restricted export opportunities for the United States' less protected and less subsidized agricultural industry. American leverage to force modifications of the European agricultural protectionism has been applied unsuccessfully for more than two decades but in recent years has been increased because of the growth of protectionist pressures in Congress and the continuing effects of losses of export shares of the Community market during the period of dollar appreciation under the first Reagan administration. For the Community, the Common Agricultural Policy, which is the official basis for farm subsidization and protection, is a major source of political cohesion, despite its high costs and unequally spread benefits, and will remain a potent unifying factor even after the completion of full market integration, especially because of its significance for the French administration, as France is the main beneficiary from this policy. Community governments are unwilling to accommodate with Americans' demands for reductions of their protectionist measures, but have begun to lower their agricultural subsidies which will diminish the volume of Community farm produce that is normally dumped on international markets. [17]

Atlantic trade in manufactured products has been influenced very strongly by the dollar appreciation during the first half of the 1980s, and, continuously over a longer period, by European subsidization of basic industries, including steel and textiles, in response to the slackening of growth in the Community since the early 1970s. Dollar depreciation in recent years has eliminated substantial surpluses in Community trade with the United States, but the United States is continuing efforts, which have been under way for more than a decade, to secure reductions of the European subsidies. The main items imported from the Community are motor vehicles (about one-eighth of the total) and petroleum (about 5 per cent). There is considerable intraindustry commerce in engines for vehicles and aircraft, and the major items for which the Community is dependent on the United States are data-processing machines and electrical gear. There is trade each way in aircraft, mostly in the United States' favour. [18]

Policy interdependencies related to the management of structural inter-dependencies have evolved in the Community's relations with the United States as a group and in the interactions of Community members with that country. On agricultural trade issues the Community as a whole deals with the United States, and this is advantageous in so far as it fosters cohesion within the Community, so as to maximize its bargaining power, although the Common Agricultural Policy is a source of intraCommunity cleavages, especially between Britain and France. On trade in manufactured products the Community also deals with the United States as a group, but member states interact directly with America as well, especially on non-tariff questions. In addition, on macromanagement issues, which affect trade and monetary relations, West Germany, France, and Britain interact individually with the United States.

Initiatives are taken principally by the United States, as the state whose commercial interests are adversely affected by the Common Agricultural Policy, the Community's industrial subsidies and the restraint in West German fiscal and monetary policies which in effect becomes obligatory for the other member states. On the agricultural trade questions the United States tends to resort to strong pressures, restricting and threatening to further restrict access to its market, but these increase France's incentives to maintain strong links with West Germany, while making all Community members aware of the politically unifying effect of their Common Agricultural Policy, despite its deficiencies, which of course are seen as matters to be settled within the Community rather than in the context of Atlantic relations. On trade in manufactured products, which involves much intraindustry commerce between the United States and the Community, more symmetrical interdependence obligates more restrained use of American bargaining leverage.[19] The American administration's inclinations to exert leverage, moreover, are being moderated as Atlantic trade moves more into balance and Congressional pressures are reduced. Current improvement in the trade balance with the Community also makes it less important to press for more expansionary policies in West Germany. In that relationship the United States has little scope to utilize any of the dissatisfactions of other Community members with West German fiscal and monetary restraint because American concern is to increase West German demand for American rather than other European products. Britain's fiscal and monetary restraints tend to preclude support for American encouragement of West German expansion, and French interest in encouraging such expansion has to be expressed with discretion in the context of the special relationship with Bonn rather than in association with the United States.

As the Community moves toward full market integration the United States' management of the relationships with the organization as a whole and with its major members may be influenced increasingly by desires to avoid provoking Community adoption of discriminatory trade, foreign direct investment and industrial measures that could affect Atlantic trade and the operations of American firms in Europe. For the Community, advances toward full market integration raise urgent questions about the extent to which the intended benefits will be secured by national firms in the member countries or by American enterprises. There is no Community consensus on methods of dealing with these issues, but one could develop if American attitudes in the trading relationship became more demanding.[20]

Community members do not regard their industrial subsidies as negotiable factors in Atlantic relations, as these aids for national firms tend to be seen as vital factors in the unequal competition against American enterprises, and their continuation is essential for the political interests of parties in office. While the forms of industrial assistance are maintained they tend to obligate arrangements for sectorally managed trade, although more with Japan than with the United States. Community firms whose competitiveness is increased have to accept restraints on their exports, as has been evident in American restrictions on imports of European steel. Community firms receiving industrial aid and lacking competitiveness tend to be protected by restraints on imports into the Community, and these have been directed mainly against Japanese products.

The lack of cohesion within the Community, its slow and conflicted decision processes and the difficulties of instituting a comprehensive common external trade policy cause reluctance to respond to American requests and to seek understandings that could lead to more harmonious Atlantic relations. In the typical sequences of interaction American pressures produce reactions which help to make possible responses that are relatively more unified and more positive in the Community interest than would be possible if members of the grouping were acting without the challenge of American leverage. The collective policy style of the Community does not have a strong anticipatory or forward orientation, and there is an incentive to defer negotiations on major substantive issues in Atlantic relations until Community bargaining power is enhanced by the growth that is expected to result from complete internal market integration.

Perspectives on the current problems of the American economy appear to reinforce Community preferences for a defensive and reactive role in Atlantic relations that can be undertaken with more unity than a promotional role. Increased cooperation from major trading partners, it seems to be recognized, might only cause American decision-makers to further postpone action on their heavy budget deficit, while remaining committed to fully autonomous management of their monetary relations, on the basis of domestic considerations. Aggravation of the American economy's problems, it may well be understood, will have to continue until the adoption of remedies is forced by a crisis.

Interdependences with Japan

The European Community members have modest asymmetric structural interdependencies with Japan which have developed principally through trade but which are being changed by Japanese direct investment. This investment, which is increasing because of the appreciation of the yen and the prospect of complete trade liberalization within the Community, has been a response to strong Community protectionism directed against Japan, provoked by Japanese informal restraints on imports from the Community.[21] After the United States Japan is the Community's main trading partner, and the commerce is moving slowly toward a balance because Japan's imports from the Community, while less than half of its exports, are increasing faster than those exports. Community imports from Japan represent about two-thirds of those from the United States, and in this commerce West Germany and Britain are the main trading partners. Both experience large unfavourable balances, while France and Italy, whose commercial policies have discriminated very severely against Japan, have maintained more balanced trade, at lower volumes.

Japanese direct investment in the Community is less than half the total in the United States, but larger than Community direct investment in Japan which is discouraged by that country's informal barriers to entry and by antipathies in the relationship caused in part by the discriminatory trade practices of some members, especially France and Italy. The Japanese outward investment is increasing transnational production interdependencies, but not to the same degree as American outward investment because the Japanese firms moving into West European operations rely heavily on sourcing from their home economy. For West European firms the Japanese presence provides technology sharing opportunities, through joint ventures and linkages, but not on the scale offered by American firms in Europe, with which more extensive cooperation has developed. If there are increasing strains in the Community's economic relations with the United States, access to Japanese technology will become more important for European enterprises.[22] Many of them, excepting a considerable number of West German companies, have to contend with technology gaps which separate them from American and Japanese firms and which are causing cumulative declines in relative productivity.

Policy interdependencies in Community relations with Japan have much less prominence than those in the Atlantic relationship and are managed more unilaterally by the Community as a whole and its individual members because of the bargaining power derived from the size of the Community market and from Japan's relative isolation in the international political economy. Ethnic and cultural antipathies toward Japan influence the policies of several Community members, especially France, and tend to prevent recognition of potentials for mutually rewarding cooperation. There is little perception of monetary and financial interdependencies, although both the Community and Japan have common interests that are vitally affected by the ways in which American policies and the performance of the American economy can impose strain on the international monetary and financial systems. The major Community members give little encouragement to Japan to assume a more active and more constructive role in its relations with Western Europe, or in the context of trilateral relations.

Community foreign direct investment policies are sufficiently liberal toward Japanese firms to facilitate steady expansion of their presence, which is beginning to involve substantial participation in the development of Community financial markets. The most important policy interactions thus continue to relate to trade, and in this area little change is occurring. Japanese firms are able to expand their penetration of the West German and British markets, as well as those of several other members, while coping with discriminatory Community trade policies which can be moderated indirectly by increasing imports from the Community and by undertaking more direct investment in the Community.[23] More meaningful and more constructive exchanges with Japan on trade policy are precluded by the unfavourable attitudes of Community governments, but some Community policy learning may gradually help to change those attitudes.

The extensive trade discrimination against Japan, mostly through the imposition of 'voluntary' export restraints, has several costs for Community members. Consumers in the Community pay higher costs for products produced by European and American firms within the groupings that are protected against Japanese competition. The main beneficiaries are West German and American enterprises, and the degrees of effective protection limit the incentives for those firms to increase their efficiency. Japanese firms, meanwhile, are challenged to increase their competitive advantages by moving more rapidly into production at very advanced levels of technology, so as to continue penetration of the Community and secure larger shares of world markets at the expense of the less efficient European enterprises.[24] The technological advances, moreover, facilitate the development of linkages with American firms in global markets, and connections with those firms can aid the growth of production-sharing arrangements with their subsidiaries in the Community to the disadvantage of weaker and less innovative Community firms. Finally, the political effects of abrasive Community behaviour toward Japan oblige reliance on the Community's own bargaining power in its relations with the United States, and there is a danger that over the long term this will weaken because of the extent to which the growth effects of full market integration will be appropriated by American enterprises in Europe rather than by the firms of Community members.

If there are severe strains in the American economy, because of failures to act on the fiscal deficits, the negative effects in Atlantic relations will challenge the Community to seek ways of strengthening relations with Japan, as well as with the rapidly growing East-Asian Newly Industrializing Countries (NICs) whose economies are becoming more closely linked with Japan. The development of wider trade and transnational production links with Japan and its high-growth neighbours, with symmetries that could be achieved through integrative cooperation, would enable the Community to cope with the destabilizing consequences of disorders in the international political economy that may be caused by a depression in the United States and a general loss of confidence in the dollar.

Trilateral interactions

The evolution of large asymmetric structural interdependencies with the United States through the expansion of the American corporate presence in Western Europe is the main feature in the emerging pattern of the Community's transnational linkages. While it continues, in the favourable context of liberal West European foreign direct investment policies and of continuing market integration within the Community, it poses increasingly significant issues for administrations in the member states regarding the effects on national economic structures and on the spread of benefits from regional integration.[25]

Engagement with these issues, however, tends to be neglected by most of the member governments, mainly because of their concerns with short-term matters but also because their doubts about the feasibility of common industrial and foreign direct investment policies incline them to see incoming American firms as desirable partners for their national firms in competition against other Community states. Significant limitations are thus imposed on the scope for collective interaction with the United States on the growth of Atlantic transnational production linkages. Member governments give their attention to the smaller but politically more prominent questions of Atlantic trade which are within the Community's formal area of competence but are still viewed as national responsibilities. On the monetary, fiscal, and financial issues which affect trade and transnational production, the Community's lack of formal competence further restricts the scope for Atlantic interaction, and this is limited at the national level because in these areas only West Germany, France, and Britain relate significantly to the United States, as individual members of the Group of Five rather than as leading members of the Community.

The contrasts regarding levels and areas of interaction are evident in Community relations with Japan, but with less significant consequences. Because the structural interdependencies with Japan are on a smaller scale, and are seen to have less significant potential, Community members give relatively little attention to Japan. In the external trade area the Community's competence is exercised to a large extent unilaterally to restrict the entry of Japanese products, and, while the larger Community members act in the same way, there is little interest in combining the Euro-Japanese interactions with those in the Atlantic context. There may well be apprehensions that in a triangular pattern of interactions Japan would tend to align with the United States because of greater dependence on the American market, and that the United States would favour Japan because of more dynamic interdependence with that state. The Community, then, could be disadvantaged. Yet current Community attitudes and behaviour give Japan incentives to strengthen political and economic ties with the United States. Prospects for the growth of these links are uncertain because of the powerful neomercantilist thrust in Japanese policies and the pressures in the American political economy to push adjustment burdens on to major trading partners. But if United States–Japan ties are strengthened, as may happen because of increased Japanese monetary and financial cooperation, the United States will have much greater bargaining power in dealing with the Community. This bargaining strength, moreover, may well be acquired while American firms in Europe are securing many of the benefits from intrazonal market integration.

The Community as a whole and its individual members in the Group of Five have little influence on the large structural and policy interdependencies which are growing rapidly in the United States–Japan relationship. Trade and transnational production links in that relationship are expanding in response to trends in the two national markets and the two differing but compatible national policy mixes, and in response to opportunities for joint ventures and linkages that serve global market ambitions. [26] European corporate involvement is overall not sufficiently competitive in global markets, and thus tends to be displaced. Atlantic policy interdependencies, as managed from the European side, affect the evolution of United States–Japan economic ties only to the extent of attracting to the Community direct American investment which would otherwise go into Japan or to the East-Asian NICs that are being increasingly linked with Japan.

Because there is little, effective interaction between Community members and Japan, the United States' use of bargaining leverage with the Community does not have to consider possible forms of Euro–Japanese cooperation which could enable the Community to assert its interests more firmly. American dealings with Japan, moreover, can be quite demanding because of that country's lack of ties with other major industrialized democracies. In this setting American administrations are not inclined to encourage the growth of Euro–Japanese links since the cost would be an overall weakening of American bargaining capabilities. The Japanese administration, meanwhile, although it may recognize the potential benefit of stronger links with the Community, has to consider not only the discouraging attitudes of Community members but also the strains that could be imposed on relations with the United States if there were an active quest for closer connections with the major West European states.

The Community, accordingly, has the most significant opportunities in the trilateral pattern. This pattern can be made more genuinely triangular, and potentially more productive, if the Community seeks wider and more active cooperation with Japan. With such cooperation, both the Community and Japan would be able to bargain more equally with the United States, while obliging the American government to accept significant degrees of external accountability for its economic policies. Of course Euro–Japanese cooperation on trade and direct investment would not readily be extended into a joint policy on relations with the United States, but shared concerns about the international consequences of American fiscal and monetary policies could well have a beneficial influence on the management of those policies. The degrees of rapport between the Community and Japan on mutual and global interests moreover would tend to make American officials sensitive to common Euro–Japanese concerns about their trading and investment relations with the United States.

Euro–Japanese cooperation could become extensive, for the development of more symmetrical and larger structural interdependencies, if sufficient cohesion evolved within the Community and if it introduced a common industrial policy that could guide harmonious integration of a substantial Japanese corporate presence into the Community political economy. Without greater internal unity and in the absence of a common industrial policy, improved political ties with Japan and greater openness to its commerce and direct investment would leave the way open for the development of very asymmetrical trade and transnational production links, and probably also for wide-ranging collaboration between Japanese and American firms in the Community, on the basis of global linkages, which would intensify competitive pressures on Community enterprises. Despite these likely disadvantages, however, improved Euro–Japanese links unaccompanied by increased political integration within the Community would tend to enhance the bargaining capabilities of both the Community and Japan in relation to the United States.

The Community option of working for a more cooperative relationship with Japan has potential significance outside the trilateral context because the growth prospects of member countries in the integrating Community market would be significantly improved by the development of large active economic links with the dynamic area of East Asian market-economy states. Stronger ties with Japan would assist the building of trade and investment connections with the other dynamic East Asian states, and, while this would ensure some protection against the adverse consequences of serious problems in the American economy, it would enable the Community to play an influential role in the establishment of a system for regional economic cooperation in the Pacific. In the long term, wide-ranging collaboration with a Pacific economic community could become possible. A free trade agreement which may be negotiated between the United States and Japan within or outside such a Pacific cooperation arrangement would of course tend to be more sensitive to European concerns if Euro-Japanese links had been strengthened.

Prospects

The European Community's management of its trilateral interdependencies is a process of insufficiently collective and comprehensive decision-making in which confederal interaction is strongly influenced by national concerns rather than the regional interest. Needs for institutional integration and for broader policy cooperation within the grouping are not being met. Because of its problems of development as an international political economy the Community is not likely to engage adequately with the tasks of managing its integrating market, for improvement of the performance of its firms and the evolution of more symmetrical structural interdependencies with the United States and Japan. The logic of advancing toward a stronger confederal or federal European system and toward monetary union is being resisted in varying degrees, especially by Britain. Market integration is thus continuing very much on a *laissez-faire* basis. Competition between Community firms is being intensified, and increasing mergers, acquisitions, joint ventures and linkages between those firms are increasing industrial concentration with mixed consequences for the multicountry Community economy. The expanding firms attain greater efficiency through economies of scale, the rationalization of their cross-border operations and faster technological advances, while less efficient firms are forced into decline. American international enterprises operating in the Community, however, mostly have larger resources for the support of competitive regional and global activities and benefit from inducements offered by Community governments as well as from the interest of many Community firms in joint ventures and linkages. West German firms are able to remain competitive, because of high levels of efficiency and the advantages of operating out of a more integrated and more supportive home environment, but other Community firms tend to be disadvantaged.

Progress toward institutional integration is likely to remain difficult for the Community because of the lack of support from member states, including Britain, which fear that their interests would be inadequately represented, but also because of the preferences of West Germany and France, whose influence would remain strong but who could have less scope to make it effective in a more representative collective decision-making system with wider powers. In the present context West Germany cannot assert strong political leadership without risking adverse French reactions, and political influence based on West German economic domination of the Community has to be exercised in the partnership with France which is a source of status and economic benefits for that country. The partnership can provide impetus for the building of a stronger system of collective management, if French policy becomes more supportive of the concept of an economic union, but in such a system West Germany could be under pressure from a coalition of economically weaker states, and this could also be a problem for France since the benefits of retaining the partnership with West Germany would remain substantial.

Because of the difficulties of building a stronger system for Community decisionmaking, issues of trilateral interdependence are being neglected, with consequences that tend to benefit outside firms and that weaken the Community's role in the international political economy while in effect encouraging Japan to work toward more cooperative relations with the United States. The need for a more actively constructive Community role in the international political economy however is increasing, on account of the growing threat of strains resulting from the United States' fiscal problems and their effects on the international monetary system.

The evolution of trilateral structural interdependencies is likely to be shaped more and more by the independent decisions of American firms, and, to a lesser extent, of larger Community firms, while the dominant role of West German enterprises in the Community is to a large extent preserved, in part through linkages with American enterprises. Attempts to evolve a common Community industrial policy are likely to be reactive, modest, disjointed and slow. Over time, moreover, the expansion of the American corporate presence through acquisitions and mergers as well as joint ventures will complicate patterns of interest representation within member–states and the processes of their policy communities to degrees that will make the evolution of a Community consensus on industrial policy very difficult to achieve.

Challenges to develop a strong system of political cooperation, to concert the policies of Community members toward outside states, will increase as vulnerabilities to the effects of the United States' economic problems become more evident and as the evolution of Community economic structures is influenced more by American firms.

The present system of consultations for political cooperation, however, is not likely to evolve into a structure for collective management of the Community's external relations unless there is considerable progress toward institutional integration in response to the tasks of promoting growth in the integrated market. The larger member states remain unwilling to lose any elements of foreign policy autonomy, and, in the West German case, this has special significance for the relationship with East Germany which is assuming greater prominence because of changes in the Soviet Union and the East European states.

Policy committees in the member states could become more aware of the need for a stronger collective decision-making structure if the European Commission were to sponsor studies of issues affecting the external relations of the integrated market. The Commission's sponsorship of research on barriers to commerce within the grouping prepared the way for the present drive to complete market integration, but left the external aspects of intrazonal trade liberalization unexplored. Policy research institutes within the Community could help to encourage Commission initiatives and could support them by organizing conferences of officials, academics and business groups.

The European Community's needs for deepening integration could also be aided, indirectly, by efforts to sponsor the development of a very active system of trilateral consultations. Such a system is urgently needed to foster cooperation between the United States, the Community and Japan for more harmonious and less asymmetrical evolution of their structural interdependencies. European efforts to establish a trilateral consultative arrangement would require new initiatives to develop rapport with Japan, and, while these would accord with the Community's own interest in promoting more triangular interactions, they would open up possibilities for a consensus on principles of domestic and international political economy management, in part because of shared problems in relations with the United States. American representatives, through involvement in the consultations, would have to respond to shared European and Japanese concerns that at present can be treated almost exclusively in separate interactions with the Community and Japan. While the consultations could be viewed as opportunities to further American interests by discouraging cooperation between the Community and Japan, the potential benefits of constructive participation could have a strong appeal, and could become more evident with the policy learning that could develop in the consultations.

Japan, it must be stressed, will probably not take initiatives for the development of a trilateral consultative system, and indeed can expect its interests to be served by broadening cooperation with the United States and continued penetration of the Community market through manufacturing for export at very advanced technological levels. Yet Japan will be able to benefit significantly from the enhanced bargaining strength derived from more active links with the Community, and from opportunities to contribute to deeper understanding, by US decision makers, of the benefits to be gained by trilateral collaboration. The significance of those benefits, of course, would depend on the extent to which the Community could play a coherent role in trilateral consultations, especially by articulating principles and norms of cooperation, and the degree to which rapport could then develop between the Community and Japan.

There is wide scope for the United States to promote trilateral consultations, but the political feasibility of initiatives for that purpose is limited by pressures to maximize the use of bargaining power in separate dealings with major trading partners, and by perceptions of West Germany and Japan as unfair trading partners, as well as by doubts about the European Community's capacity to assume a sufficiently constructive role. The utility of the separate dealings of course has to be considered with reference to the disadvantages that could result from fully triangular interactions in which the United States could face demands from both the European Community and Japan and in which Japan's interest in cooperation with the United States could be expressed with greater flexibility and enhanced leverage. Because of its extraordinary need for cooperation from major trading partners to overcome its fiscal and trade problems, the United States has incentives to make full use of its negotiating strength and arguments to this effect have potent influence on Congressional attitudes. The logic of working for an active triangular consultative arrangement *in the general interest*, however, may be given more and more expression in American policy if a sufficiently vigorous transnational network of officials, business leaders and scholars is organized for the promotion of trilateral policy learning. Some inspiration for the development of such a network can be derived from Paul Volcker's remarks, at the Tokyo meeting of the Trilateral Commission last year: 'The more diffuse world near the year 2000 is going to require a more structured international system than we have now. It's going to require a stronger GATT. It's going to require a structured monetary system . . .' These remarks followed observations on the need for international leadership and were followed by comments to the effect that the leadership functions of the trilateral grouping would have to be shared more widely as the more structured global economy emerged.

References

1. See John Zysman, *Governments, Markets and Growth*, Ithaca, Cornell University Press, 1983, ch. 5; Simon Bulmer and William Paterson, *The Federal Republic of Germany and the European Community*, London; Allen & Unwin, 1987 and Russell J. Dalton, *Politics in West Germany*, Glenview, Scott, Foresman, 1989.
2. See 'Symposium on Long Run Economic Performance in the UK, *Oxford Review of Economic Policy*, 4, 1, Spring 1988 and R.E. Rowthorn and J.R. Wells, *Deindustrialisation and Foreign Trade*, Cambridge, Cambridge University Press, 1987.
3. See *Structural Adjustment and Economic Performance*, Paris, OECD, 1987.
4. See Alberto Giovannini and Francesco Giavazzi, *Limiting Exchange Rate Flexibility: The European Monetary System*, Boston, MIT Press, 1989.
5. See observations on the sequences of events in Pierre-Henri Laurent, 'The European Community: Twelve Becoming One', *Current History*, 87, 532, 1988, pp. 357–60. See also *Efficiency, Stability and Equity* report of a study group for the Commission, Brussels, European Commission, 1987.
6. See James M. Markham, 'Europe, Facing Tough Decisions, Now Cautious on a Single Market', *New York Times*, 11 March 1989.
7. See *Survey of Current Business*, 68, 8 August 1988.
8. See Gavin Boyd, *Pacific Trade, Investment and Politics*, London, Pinter Publishers, and New York, St Martin's Press, 1988, ch. 7.
9. See general observations on this process in John Cantwell, 'The Reorganization of European Industries after Integration: Selected Evidence on the Role of Multinational Enterprise Activities', *Journal of Common Market Studies*, XXVI, 2 Dec 1987, pp 127–52. See also Graham Hall (ed.), *European Industrial Policy*, New York, St Martin's Press, 1986.
10. See Jacques Pelkmans, 'Economic Relations between the European Community and East Asia: Protectionism or Cooperation?' in Philip West and Frans A.M. Alting von Geusau (eds), *The Pacific Rim and the Western World*, Boulder, Westview Press, 1987 pp. 201–36.
11. See Giovannini and Giavazzi, op. cit.
12. See observations on interdependencies in Robert E. Baldwin, Carl B. Hamilton and Andre Sapir (eds), *Issues in US-EC Trade Relations*, Chicago, University of Chicago Press, 1988.
13. Ibid, and see *Survey of Current Business*, cited.
14. On the dominant roles of American international firms in the world economy see Jeremy Clegg, *Multinational Enterprise and World Competition*, New York, St Martin's Press, 1987.
15. Ibid.
16. See *Survey of Current Business*, cited.

17. See discussion of agricultural trade in Baldwin *et al.*, op. cit., part 2.
18. Ibid, part 4.
19. Ibid.
20. See Michael Calingaert, *The 1992 Challenge from Europe*, Washington DC, National Planning Association, 1988.
21. See Gerald and Victoria Curzon, 'Follies in European Trade Relations with Japan', *The World Economy*, 10, 2 June 1987, pp. 155–76.
22. See Frank Press, Hubert Curien, Carlo de Benedetti and Keichi Oshima, *A High Technology Gap: Europe, America, and Japan*, New York, Council on Foreign Relations, 1987. On Japanese direct investment in Europe see statistics in Yasusuke Murakami and Yutaka Kosai (eds), *Japan in the Global Community*, Tokyo, University of Tokyo Press, 1986.
23. See Gerald and Victoria Curzon, op. cit.
24. Ibid.
25. See general discussion in John H. Dunning and Peter Robson, 'Multinational Corporate Integration and Regional Economic Integration', *Journal of Common Market Studies*, **XXVI**, 2 Dec 1987, pp103–26.
26. See Thomas K. McCraw (ed.), *America Versus Japan*, Boston, Harvard Business School Press, 1986 and Farok J. Contractor and Peter Lorange (eds), *Cooperative Strategies in International Business*, Lexington, Mass, D.C. Heath, 1988.

9 The American View of EC-1992

Gregory J. O'Connor

Introduction

One of the most important events facing American companies involved in international trade is the European Community's programme to create a single market by 1992. Important things are happening in Europe today. The European Community is engaged in a fast-track programme to remove barriers to the free movement of goods, people and services, to form a single continental market. By 1992, the European community expects to do away with most of the non-tariff barriers that restrain trade within their own market. If successful, this will result in a $4 trillion market of 320 million consumers. These changes will require a strategic response from business, as well as a technical restructuring, to take account of the EC's single market. I would like to review the important aspects of the 1992 programme highlighting the important opportunities and challenges for American business. I will also touch on the more important commercial implications and finally summarize the strategic concerns for American business regarding the competition in a single European market.

In this chapter, I would like to review the important aspects of the 1992 programme highlighting the important opportunities and challenges for American business. I will also touch on the more important commercial implications and finally summarize the strategic concerns for American business regarding the competition in a single European market.

I want to call your attention to the three major factors that not only will affect the technical nature of doing business in Europe but also help determine the strategy that can be used to approach Europe in 1992. The EC's efforts to create a single market is first of all a broad array of technical, regulatory and legal changes in the business environment. The programme includes (at most recent count) 279 directives for 1992 to harmonize the EC business environment and eliminate barriers to internal European trade. These technical changes are far-reaching and have strategic implications for American business in Europe and for those exporting to Europe.

These changes will affect every area of commercial activity from the introduction of a common customs document which eliminates the need for different customs forms in the EC countries to harmonization of product standards, to liberalization of capital movements and a harmonized policy for mergers and acquisitions. Any one of these 279 directives requires a planned response from affected American business executives. Together these directives require a response from all business executives.

The second part of the EC's programme is what happens after 1992. The European Community has already begun to consider ways to create a common currency and a common banking authority. Most Europeans are convinced that the commercial benefits of the single market can only be obtained if a more integrated financial market is developed in the Community. The recently issued Delors Committee report on economic and monetary union signifies the importance the EC has attached to this issue. The result of all these changes is that the competitive environment will be fundamentally altered. Closed markets will be opened up. If a product can be sold in one EC country it can be sold in all. An establishment in one country can make sales throughout the community.

Finally, 1992 also means that the European private sector will respond to these changes with new business structure, new distribution systems, new computerized support services, new business alliances. This is a particularly important area for American business people to follow. Member-state governments are also working now to assist their small and medium-sized businesses adjust to a more competitive commercial environment. The result will be tougher competition and the emergence of stronger and larger European firms that will challenge American and other firms world-wide. An example of this relatively new phenomenon is Grand Metropolitan's purchase of Pillsbury.

The opportunities for American companies

American companies should have important opportunities to increase their sales to Europe as a result of the 1992 programme. The European community should experience faster economic growth—which will increase the demand for American products in the EC. Second, it will be an easier and more attractive market in which to sell. Procedures and standards will be more uniform, reducing many of the costs and difficulties that have characterized exporting to the EC countries.

One of the helpful achievements to date, for example, has been the implementation of 'the single administrative document'—a single customs document that has replaced up to 150 documents for some export shipments in Europe and has greatly reduced the cost and administrative burden of shipping goods throughout the EC. This is of particular benefit to smaller companies. Another helpful development is the effort to eliminate standards as an internal trade barrier. This is of particular interest to smaller American companies unable or unwilling to make up to twelve variants of one product to meet the standards requirements of the twelve markets comprising the EC. Many smaller American companies export to only one or two of the largest EC markets, and the cost of meeting different standards has been one of the reasons why they have not attempted to sell more broadly in the EC. As internal barriers are eliminated, companies doing business in and with Europe will be able to achieve greater economies of scale. Plants can be rationalized and expanded in Europe and marketing and exporting costs will be reduced significantly.

Observers believe that companies can reduce a large share of their overhead costs due to the liberalization of transportation regulations alone. Inventories as well as distribution systems can be managed much more efficiently. One of the questions we in the commerce department get from smaller American firms is, 'will I have to invest in the EC to keep selling there?' The answer to that is 'no'. Tariffs are not being raised, and external barriers—with some possible exceptions I will discuss in a few minutes—are not being raised. It is true, however, that the reduction in internal barriers will raise the amount of competition encountered within the EC.

Potential problems for American companies

By now most people are familiar with the statements of EC officials that Europe 1992 will not be a 'Fortress Europe' but rather will be a 'world partner'. The United States supports a European programme dedicated to opening markets and expanding trade and investment opportunities for the world economy. Over the last several months, our office has spoken to hundreds of American business executives about 1992. Everyone sees the opportunities. But it is also true that there are still questions remaining about the impact of the programme on American trade interests.

Together, the EC and the United States produce two-thirds of the Western world's GNP. Europe is our largest customer and our largest competitor. If this single internal market unification results in more open markets and promotes growth and innovation, it will be good for American business. However, if the EC programme results in compromise of free-market principles, or in greater protectionism, neither American nor Europe industry will benefit from these opportunities. And our largest foreign market could be threatened.

Department of commerce specialists have been analysing the commercial effects of many of the proposed 1992 regulations. For the past several months, we have been working closely with over 100 trade associations and American companies to prepare these analyses. It is clear that for the most part the new EC regulations are not discriminatory, but in some areas they may increase the preference for European products made by European companies. A good example of this is in the EC's process of harmonizing their product standards, one of the most important aspects of the 1992 programme, and potentially of the greatest benefit to American companies doing business in Europe.

We must first distinguish between 'essential requirements' and 'standards'. *Requirements* set a general framework and they are voted on in the regular EC legislative process. An example is the machine safety directive. This provides broad guidelines concerning the safe use of electrical machinery. *Standards* are far more specific than general requirements. The EC does not develop standards; this job is delegated to CEN/CENELEC, a private standards-setting body with a membership of governmental and private standards bodies from the EC member–states and the EFTA countries. CEN/CENELEC is closed to outside observers.

We are concerned that the European Community is not allowing enough transparency into their system. It is presently difficult and time consuming for American companies to know what is going on in the EC process of drafting product standards and even more difficult to make their comments and concerns known to the EC. We are urging the Community to implement a programme that will allow participation of interested American companies in standards drafting and in certification and testing systems. Despite the fact that these new standards set essential safety and health requirements, local preference remains an important consideration. For example, the West German beer purity law effectively kept out imported beers because it did not allow any preservatives to be used. Several member states challenged the law as a non-tariff barrier to trade and it was was struck down by the European Court of Justice. However, many, German brewers still label their beers as being brewed in accordance with the age-old law that has been officially overturned.

Another area of concern is the EC proposal to open up public procurement in previously excluded sectors of telecommunications, water, transport and energy. American companies located in Europe stand to gain from liberalization in this area. However, these opportunities would be compromised by the Community's present proposal which calls for a 50 per cent local content restriction and, in some cases, a further price differential for EC bids.

We also continue to be concerned that the EC may attempt to use the criteria of 'reciprocity' before granting improved market access to non-EC companies. This has been mentioned most frequently in the context of banking and financial services, sectors that are not covered under the present international trading rules of the GATT. The United States government has emphasized our concern with this concept of reciprocity, and the EC is currently re-thinking their proposal. We welcome the EC Commission's revision to their reciprocity provision in the second banking directive.

The EC has adopted a position that national treatment should be the standard for a reciprocity provision. Since the United States already gives national treatment to EC banks, and allows them comparable access and competitive opportunities, we strongly believe that American banks should not be in any way discriminated against by this new reciprocity clause. But this is an area where we will need to continue to be vigilant in protecting the right of American companies to national treatment in their European activities.

Another area of concern is that the EC will introduce so-called 'transitional measures'—temporary protection to allow European companies time to adjust to more open competition. Some European officials, for example, are talking about shielding European auto companies from foreign competition. At present, they are searching for a way to incorporate nationally administered quotas on Japanese cars (in Spain, Portugal, France, Italy and the United Kingdom) into a single EC-wide quota. (Example: this divisive issue is exemplified by the case of the Nissan Bluebird manafactured in the United Kingdom. The British want to ship the car to the continent. The French initially insisted that the cars be considered Japanese and counted against the quota because of their limited amount of European local content. However, the French have recently stated that they will begin accepting the cars as British beginning at the end of the year. The Italians, on the other hand, adamantly refuse to accept the cars as British.) There is strong resistance from several automakers with protected home markets because these 'national champions' stand to lose from a more competitive marketplace. Those industries that will be negatively affected by the single internal market can be expected to resist and try to prevent the programme from being fully implemented.

The outright barriers to trade so far are few that are directly related to 1992, but 1992 initiatives can mean increased costs and these costs could affect American businesses more significantly than competitors in Europe. Let me just set out a few examples of what I mean. For the most part it is still too early to identify a definite direction toward an industrial policy in technology, but recently the European Community Commission has adopted origin rules for semiconductors 'and in contemplating local content rules for various sectors that should cause American firms to look at their investment in Europe and other parts of the world. Examples of recent investment include Fujitsu's announcement that they will build a semiconductor chip plant in northern England.

Requirements in product testing and certification under the 1992 rules for harmonizing product standards present a preference for European third-party testing and certification. Although not discriminatory, the effect of these requirements could well be to draw more and more product testing, product development and even product design on shore in Europe so that test data and procedures are more readily available to support regulatory and legal requirements.

The European community is becoming very interested in the social dimension of the single market. Some of the 1983 proposals for mandatory work councils and worker consultation and co-determination are being reconsidered in the context of the social and labour effects of the business restructuring that is anticipated as a result of the single market. Some businesses in Europe see such rules as inevitable and are working to include increased flexibility and competitiveness in the response to restructuring.

Environmental issues are becoming much more important in Europe and the single market is driving consideration of EC-wide harmonization as an important objective. In July, for example, the EC Commission proposed directives that would establish EC-wide rules for handling solid waste and hazardous waste issues. There are concerns over this in that drafts of regulations could be interpreted to require the producer and all subsequent holders of waste to be liable for final disposal costs.

Competition rules on exclusive distributor agreements should remain pretty much as they are today. In fact, a single trademark and patent may strengthen the commercial value of these rules. But the EC Commission is re-examining exclusive networks and agreements in certain sectors (pharmaceuticals) and this could affect one of the most important marketing tools for American exporters.

What should American companies do?

The question remains, what should American industry do? First, it seems clear that your export, versus investment, decision should still be an open one, based on normal commercial criteria. It still makes good commercial sense to start penetrating the EC market with exports. It is important to realize that the end of 1992 is a deadline for action. The EC's programme to unify the internal market is being implemented now and through 1992. By 1992, the EC market will have already changed and will have fundamentally affected the business prospects of those who are ready and those who are not.

American companies have an important and challenging job ahead in preparing themselves for the technical and strategic changes they will face in Europe. I can state unequivocally that the United States' government is committed to working closely with the business and academic community to assure that American industry realizes the full potential of the 1992 programme. U.S companies should target the 1992 programme as a major priority for action, and begin to organize themselves effectively to take advantage of the changing European market-place as well as to be able to deal with the potential threats. That means, first of all, becoming familiar with the details of the 1992 regulations and determining how they will affect their interests. American firms need to understand whether their products could now meet the proposed new standards and analyse how their operations in Europe should be strengthened to take account of a new European environment. American businesses need to know what their competitors in Europe are doing and how this might affect their present position and market share.

As I said at the beginning 1992 is more than a technical programme. It should also prove to be a challenge to business strategies. The American response cannot be limited to a technical response. The drive to create a single market was among other things a response of Europe to the growing international competition and the growth of a world economy. Pressures in the global market make Europe a new battleground for world markets. The single market is attracting Japanese and EFTA country investment as they seek to position themselves better in a new and potentially lucrative market—the EC's single market (Swedes are getting nervous). This means that for firms already used to doing business in Europe competition could be tougher.

Most of the merger, acquisition, and joint-venture activity in Europe reflects a strategic response to these factors. In fact a great many American companies have increased their interest in investment in Europe for a variety of reasons. Many firms have realized that asset prices are rising very rapidly in the EC and that now is the time to purchase assets in Europe. Also many American business are preparing for the single market particularly by developing cross-border distribution channels. Since a great many European continental businesses are privately held this more often follows the joint-venture route. Examples of this kind of behaviour include: the joint ventures between General Electric (United States) and General Electric Corporation (United Kingdom); the joint venture between Whirlpool and Philips of the Netherlands in the home appliances area and AT&T's joint venture with Italtel in the field of telecommunications equipment.

Another very important factor is that Europeans are preparing their own strategy for competition in 1992. A great deal of the European investment in 1992 and in American markets reflects their plan to be a more competitive force in world markets. European companies themselves are making investments in Europe and globally with the objective of becoming the lowest-cost producer in their market niche and garnering the largest market share for their products. The American business community needs to respond to these strategic changes as well as the technical changes precipitated by 1992. American industry should be seeking new opportunities to exchange views with European industry on how to respond to the changing European environment. A constructive dialogue between American and European industry will be beneficial to all concerned.

The United States' government role

In all of these activities, American industry can count on the assistance of the United States government. The commerce department has established the EC 1992 programme as a top priority. We are committed to helping American companies respond to the challenges and opportunities of 1992. We have established a 'single internal market information service' to help inform the American business community about the details of the EC's programme, to help solve problems where they occur and to assist American companies in taking maximum advantage of the opportunities. As of the summer of 1989 we had already provided information and advice to over 8,000 companies. United States' government officials are working actively to assure that the European

Community does not implement programme that will disadvantage American commercial interests. We are committed to seeking resolution of individual problems as they arise and to utilizing the full measure of American trade law to protect American trade interests if that becomes necessary.

Conclusion and summary

There is going to be a European single market that the American business community should prepare for. This is particularly the case for the goods-producing industries. At the outset, at least, the European market is going to be a very complicated business environment. There will be overlapping and conflicting regulations and jurisdictional issues. The market will be heavily regulated, perhaps in countries where regulations were previously minimal.

While there is a measure of transparency for the programme as it is being formulated in the EC Commission, there is no road map or experience on implementation. New agencies, policies and dispute settlement mechanisms will have to be invented to cope with the enormous task of delivering on the legislated 1992 programme. European companies are attempting to become more competitive in their own market. Commercial preferences should be valuable to EFTA companies and increase competitive pressures in Europe. Certainly, with the movement to a single market, Japan and other producers will become more interested in Europe. The European Community 1992 programme presents an exciting challenge for American industry. The time to act is now. The United States Department of Commerce stands ready to help.

10 The impact on global corporate competition and multinational corporate strategy
Lawrence G. Franko

The beginning

Perhaps the most important fact about what has become known as 'Europe 1992' is that it is the direct result of a business initiative aimed at making European industry more globally competitive. The lineage of the Single European Market can be traced to two businesses and to two businessmen in particular: Philips of the Netherlands and Volvo of Sweden, whose respective presidents are Wisse Dekker and Pehr Gyllenhammar (a view independently confirmed by Jeelof, 1989 and d'Oultremont, 1989). Gyllenhammar in turn is the chairman and original organizer of the Roundtable of European Industrialists, a group whose membership is a virtual *Who's Who* of large European multinational enterprise (*The Economist*, 4 July 1987). Although the role of European politicians and civil servants, in particular Jacques Delors and Lord Cockfield, who himself had a business background as former head of Boots, a British pharmaceutial firm, has been much more visible in press coverage of the move toward the Single European Act of 1987 and the road to 1992, the initial parental and conceptual impetus came from elsewhere. Indeed, if a politician or civil servant might be singled out when the definitive history is written, it would arguably be Vicomte Etienne Davignon, who understood that, left to their own devices, national governments were not likely to build Europe. If they were, it would only be in response to business initiatives (d'Oultremont, 1989).

The next most important facts, to wit, what triggered the pressure for 1992, and what it is supposed to accomplish, are directly related to the first. Philips is an electronics company and Volvo's primary activity is in motor vehicles. Few sectors have been under as heavy attack from the Japanese in world-wide competition, and few have been as Balkanized as these by non-tariff barriers to trade within the European Community's not-very-common-market.

The impetus for 1992 came out of a sense of loss and threat—from Japan. Back in 1968, Servan-Schreiber's *American Challenge* had pointed to American industry as the principal menace to European high-technology competitiveness. Twenty-five years later, it had become abundantly clear that the main threat was Japan. With the exceptions of aerospace and computers, Europe's high-tech hopes—semi-conductors, consumer electronics, telecommunications, motor vehicles, and office equipment—were under siege primarily from the Orient, not the United States.

European industrialists had observed events on the first field of battle for world market share in technology-intensive goods, the United States, where the Japanese had won more often than not. They did not wish for a replay, national-market-by-national-market, in Europe.

The Japanese threat was, moreover, perceived as being more dangerous than any from the United States, even in the high-tech sectors where American firms were still prominent. The Japanese appeared bent on total market domination wherever they struck. The behaviour of American firms in Europe, in contrast, was typically described in terms of a willingness to 'settle in' once a 'reasonable' market share had been obtained. Americans were viewed, in the words of the vice-chairman of Philips, as 'willing to establish a meaningful production, employment and R & D presence in Europe' (Jeelof, 1989). They were also perceived as wanting profits, sooner or later, and therefore as reluctant to engage in margin-grinding fights to the finish. With local profits came tax payments to governments, another factor easing the adjustment in Europe to any successful American challenges. Also, given the comparative ease of European entry into the American market, including entry via acquisitions, American firms were reluctant to provoke European retaliation 'behind home lines' by engaging in 'excessively ungentlemanly' competitive behaviour. In contrast, competition with the Japanese felt more often like total war.

Although the sense of threat from Asia was widely shared among European industrialists, many of whom formed a 'common front' in the form of the Roundtable presided over by Volvo's Gyllenhammar, it is worth dwelling for a moment on why Philips and Volvo—one of which is not even headquartered in the EC—might take the lead. One common denominator is that both lack a large national home market. They thus lack not only the scale advantages (and potential political clout in government-to-government trade negotiations) offered by a large home market, but they also lack *the* fundamental element of the Japanese reality which provided the springboard for the Japanese threat. (On the role of the large, homogeneous Japanese home market as 'spawning ground' for later global success, see Franko, 1983, especially pp. 35–8.) Both had seen the Japanese pick-off lower-priced segments of their product ranges, and begin preparation for an assault on 'up-market' television sets, audio equipment, cars and trucks. Both had also seen the Japanese use their cost advantages to dominate the markets of the smaller, more open European countries, and then use these markets as staging bases for an assault on larger markets—sometimes after adding on what have come to be called 'screwdriver', or final-assembly-only plants, as a means of evading the national trade protection that might blunt or delay Japanese thrusts.

For Philips, insult was arguably added to injury. It had, after all, introduced the first consumer VCR, only to find that rapid development in the 'Japanese common market' (based on Japanese-determined technical standards) had virtually eliminated it from the world market. In addition, Philips had seen almost all other significant non-Asian producers of consumer electronics in the world simply eliminated from the industry. A desire for survival in consumer electronics was coupled with a desire to avoid a defeat in the future in industrial electronics, electrical equipment and telecommunications. Philips is also one of the three or four largest single private-sector employers in the EC countries and it is almost certainly the largest such employer in the Benelux—the very core from which the EC grew. It also has employment and production facilities distributed more widely around Europe than does virtually any other major European firm. It is a European as well as global multinational, but cost-increasing non-tariff-barriers (NTBs) account for a good deal of its European 'multinationality'.

European business goals for Europe 1992 are thus to give European industry the large home market it has never had, and to lower costs in competition with the Japanese by eliminating the need for high-cost production and sourcing behind NTBs. Moreover, it bears repeating that 1992 is happening in response to challenges of *global* competition and therefore has business strategy and economic consequences which go well beyond Europe.

The future

At first blush, the implications for business, and especially for European multinational businesses, of a move to a single European market would seem to be simply those heralded back in the 1960s for the single market that was supposed to have emerged, but did not, back then. Then, the clarion calls were for the realization of scale economies through the specialization of plants having uneconomically small production runs in limited national markets, an industrial re-shuffling often termed 'rationalization' in Europe. Today, much of the rhetorical emphasis is similar, but, especially with the addition of the Iberian Peninsula and Greece to the Community, and the likelihood of closer EC relations with Turkey and Eastern Europe, the industrial and financial consequences are likely to be far wider.

Barriers to trade protect high-cost production. More accurately, trade barriers protect and benefit high-cost factors of production, to wit, skilled or unskilled labour and capital. With the unification of capital markets in the EC there should essentially be no differences across countries in the long-term costs of capital, however much certain individual small plants in, say, Spain and Portugal may prove to be uneconomic immediately after the opening of EC-wide competition. Thus, the optimal siting of production across the face of Europe will primarily follow labour-cost differentials. In plain English, high-skill high-tech may remain in the north, but the centre of gravity of low-and medium-technology production will be moving south.

Indeed, the combination of improving educational standards and the new political maturity in southern Europe, plus coming demographic stagnation in the north could lead to a dramatic re-arrangement of the European economic map during the coming decades. German trade unions and portions of the French left are understandably upset (New York Times, 8 March 1989). France is notorious in the business world as the still-most-protected market in the EC. What is less often realized is the extent to which national industry standards and governmental purchasing preferences have protected important segments of West German industry. It is no accident that the very issue that moved Europe to the principal of 'mutual recognition' of product standards among member nations, trade in *crème de cassis*, involved West Germany (*The Economist*, 'A Survey of Europe's Internal Market', 9 July 1988).

Perhaps German trade unions are not too upset, however, for maintaining traditional manufacturing in the north would require the import of increasing numbers of immigrant workers. They may only wish to slow the pace of change, and to make sure that the change to Iberia occurs in the hands of, say, Volkswagen or Siemens so that their pensions remain secure. (Pension provisions in Germany are reinvested internally in firms, as opposed to the American practice of 'funding', that is externally investing pension fund assets. Thus, pension payments are contingent on long-term company viability.)

1992 . . . When?

If the 'where from' and 'where to' aspects of 1992 are clear enough, what can be said about 'when', or, as the question is often phrased in the United States, 'can one really believe that the fractious Europeans will *really* have a "Europe without borders" by 31 December 1992?' The easy, and trivial, forecast is that they will not. It is a virtual certainty that three-and-one-half years is too short a time in which to dismantle fully all of the many national trade-distorting practices that have been built up over decades, if not centuries, among the nations making up the EC. These barriers include, after all, a myriad of national public-procurement preferences; health, safety and environmental regulations; industrial standards, testing and certification norms; transport regulations, permits, and cartels; value-added-tax differentials; national banking and financial services rules; and national professional and degree certification requirements.

What matters most from a company strategy point of view is not that the EC 'truly becomes' a single market anytime soon, but that a credible process has been set in motion which suggests that there is a real probability of such an eventuality. Such a process is clearly in motion. Indeed, the intriguing thing is that so much has happened already as of 1989, not only in the legal, political and judicial sphere in terms of EC directives and 'alignments' and 'mutual recognition' of national regulations, but also and more importantly in company behaviour.

The company dimension

Company activity in regard to '1992' has been by no means limited to lobbying in favour of, or against, measures to promote a single European market. Indeed the most visible responses to 1992 have been in the form of a seemingly bewildering number of mergers, acquisitions, divestments, restructurings and joint ventures involving not only EC corporate 'insiders', but also American and Japanese 'outsiders'. 'M & A', or mergers and acquisitions or bids therefore, and large 'deals' are of course the stuff of which business press coverage is made. They, however, are not and should not be the only corporate activity going on, if the lower-cost benefits of scale economies and relocations of production that ought to happen as a single European market are truly to occur. Indeed, lost sight of in the deal announcements presented invariably as 'positive responses to 1992' is an unfortunately long tradition of *defensive* alliances, agreements and mergers in Europe whose corporate purpose has often been to substitute private barriers to international trade for declining public barriers (Franko, 1976, especially chapters 2 and 6).

Understanding the corporate responses to Europe 1992 which have already occurred, and forecasting the possible shape of corporate strategies, requires a review of company behaviour, sector-by-sector, to which we will shortly turn. While undertaking this review, however, it may be worth bearing in mind that there are at least five possible categorizations of corporate behaviour by actor and motivation in response to 1992—and that responses may differ dramatically in consequence. The categories one can distinguish include:

1. European companies choosing supranational, *offensive* strategies, that is those strategies exploiting scale and optimizing location of activities;
2. European companies choosing international, *defensive* strategies, that is acquiring, allying with or merging with others in order to reduce or inhibit the need to compete supranationally;
3. European companies choosing *national, defensive* strategies, that is first and foremost orientated toward protecting their 'home turf' from invasion;
4. Strategies of the '*old outsiders*', that is the American multinationals who already have a significant European market and production presence;
5. Strategies of the '*new outsiders*', that is of Japanese (and other Asian) invaders, as well as of American firms without an historical presence in Europe.

Of course, to further complicate matters, groups (1), (2) and (3) may not be entirely mutually exclusive. Some European firms may hope that initial defence will provide a trampoline for later offence. Moreover, not all group (4) or (5) 'outsiders' may prove alike. Still, even a rough categorization underscores the point that company behaviour may indeed be very different depending on firms' motivation and initial competitive position.

A sectoral overview

A start at understanding the corporate strategic response to 1992 can be made by surveying the major events that have already occurred in 'the most affected industries' through mid-1989. Such industry portraits follow, with the caveat that they are neither comprehensive, nor, since company moves are an ongoing process, are they likely to be wholly up-to-date by the time the reader sees these remarks.

Telecommunications equipment

The principal events of the 1980s in Europe's telecommunications industry included:

- The sale by ITT of the United States of its European telecommunications equipment activities (which dated from the 1930s) to France's CGE/ Alcatel, following on technological failure and lack of commitment to the sector by conglomerate ITT;
- The acceptance by the French government of Sweden's Ericcson as owner of the 'French alternative supplier', CGCT—rather than ATT of the United States or Siemens of West Germany; and
- The 'difficult entry' into Europe by ATT, first by a not especially successful joint venture with Philips and more recently *via* an ownership/ alliance/ supply agreement with the Italian state-owned STET.

To these re-arrangements can be added the (as yet unconsummated) joint bid for Britain's Plessey by GEC (United Kingdom) and Siemens of West Germany, with Siemens being the interested party for the telecommunications (as opposed to the defence electronics) side of Plessey.

Commenting on ATT's problems in obtaining access to European markets and production sites, the vice chairman of Philips has remarked that 'they [Philips] underestimated the resistance of European PTT authorities to "outside" suppliers, even in the company of a European partner' (Jeelof, 1989). Even European moves across European borders have encountered ambivalence, as is suggested by British government hesitation over the GEC–Siemens bid for

Plessey. It would seem that a West German presence, and possible domination of British production is greeted with less than total enthusiasm—even as a supplier to privatized British Telecom.

Despite great strengths in technology and in Asian markets, a Japanese thrust in Europe has been conspicuous by its absence, or rather prevention as a result of the historic national purchasing preferences of the European government PTTs.

Electric power generation

The story here has been one of the creation of a European 'triopoly': of the ASEA (Sweden)–Brown Boveri (Switzerland) merger, which brings together many production and sales sites within the EC of those two EFTA-based parents; of the rise of France's Alsthom (also of the CGE group, as is Alcatel of telecommunications fame) and its recent incorporation of the power turbine activities of Britain's GEC; and of the continuing, if less visible role of Siemens of West Germany.

The American company dimension has been one of exit, not only from Europe, but from the United States—most notably in the partial sale to, and taking control by, ASEA–BBC of Westinghouse's American power generation activities.

Japan's electrical power equipment firms, notably Hitachi, and Mitsubishi Electric, have not—or have not been able to—move into this sector where European purchasing decisions have traditionally been taken by state-owned, national electricity authorities.

Consumer electronics and appliances

In consumer electronics, that is, television sets, VCRs, audio, calculators and the like, the European and indeed world shakeout provoked by the Japanese onslaught already took place during the 1970s. American firms, with the exception of Zenith, are now virtually out of the industry altogether, much to the chagrin of America advocates of high-definition television technology who have no American national champion to encourage. The European industry has shaken down essentially to three firms: Philips, Thomson of France, and (much smaller) Nokia of Finland. Thomson's takeover of the television activities of RCA and GE in the United States, had, however, an arguably '1992' rationale: that surviving in an industry driven and dominated by Japanese firms (and Philips) who had a *world-wide* scope, required global scale, volume and market feedback from the world's largest TV market, the United States. A major America presence was therefore required, if European competitive cost and technological viability were to be maintained. Interestingly, Thomson acquired RCA and GE's consumer electronics in a 'swap' with America GE, which in turn obtained Thomson's medical electronics activities, thus making GE number two in that business in Europe to Siemens. It might be noted that GE's interest in European medical electronics appeared largely motivated by prior moves by Siemens into GE's American home market.

In appliances, (partial) ownership and control divestments by Philips and GEC (United Kingdom) have provided opportunities for American entry into Europe, both by 'new multinational' Whirlpool, which took majority control of Philips's appliance division, and 'old multinational' GE (United States), into whose arms GEC (United Kingdom) put its appliance (as well as other) activities when threatened with takeover and dismemberment itself. These America ventures into Europe are, however, perhaps less pure 'opportunity seeking', than a response to thrusts into the American market by European appliance leader Electrolux of Sweden, which holds some 25 per cent of the European market and which has grown to its present dominance by acquiring and rationalizing many smaller European firms on the basis of supranational production and sourcing strategies (*The Economist*, 'Electrical Brief', 11 February 1989, p. 68).

Banking and financial services

It is often difficult to distinguish competitive from cooperative moves in an industry where correspondent, syndicate and referral relationships have a history of decades if not centuries. Although one knowledgeable observer has described the '1992 moves' of European banking as a 'phony war', (de Jonquières, *Financial Times*, 5 May 1989, p.38) in which major changes will come about from national deregulation now underway, as opposed to being motivated by cross-border, price-cutting competitive raids, supranational moves there none the less have been.

As in telecommunications and power equipment, some of the most visible moves have coincided with American withdrawals from Europe. These have been motivated in large part by American banks need to rebuild capital in order to deal with write-offs of unserviced Third World debt. The Deutsche Bank's most significant cross-border move was the acquisition in 1986 of Bank of America's former Italian subsidiary:

. . . one of only two private-sector banks in Italy with a well-developed national branch network. At the end of 1987, Bayerische Vereinsbank bought the Rome and Milan branches of the First National Bank of Chicago. Also, in December 1987, Credit Lyonnais acquired Nederlandse Credietbank, the Dutch subsidiary of Chase Manhattan . . . Last July . . . Banque Nationale de Paris purchased Chemical Bank's British home mortgage subsidiary. [J. Thornblade, 'The Impact of EC 1992 on Financial Institutions and Markets', Bank of Boston, 6 January 1989]

Cross-border 'mergers among European equals' have been lacking, but cross-ownership ties, such as those between the Bank of Scotland and Banco Santander in Spain, and the European network established by Hambros, United Kingdom, have occurred. The Dresdner Bank of Germany and the Banque Nationale de Paris have exchanged board members, if not yet ownership (*Financial Times*, 15 June 1989, p. 23.).

Cross-market 'invasions' have also been announced. Barclay's has established a European-wide branch network and plans to market mortgages, insurance and money management services across Europe.

National mergers and acquisitions, notably among the large banks in Spain, and those 'across the financial services spectrum' between banks, insurance companies and brokerage houses, especially in the United Kingdom, have also been noteworthy. Although 'horizontal' mergers among national banks may be at least partly defensively motivated, the experience of American financial deregulation suggests that 'inter-industry' invasion across the boundaries of financial services are among the most potent forces of price-reducing competition.

The 'big four' Japanese securities houses have long been present in London and Switzerland, as have Japanese banks. The latter in particular are said to be expanding their offices in West Germany, albeit for wholesale rather than retail banking, at a blistering pace. Heretofore, however, it appears that much of their activity as been in servicing Japanese trade and investment in Europe, rather than in directly competing with European banks.

In contrast to many American banks' apparent withdrawals or phase-downs of European operations, some American insurance firms, notably the American International Group, have been aggressively expanding, using Spain, with its relatively underdeveloped institutional insurance market as a prime base. European, and Japanese, insurers appear to have been at least as fast off the market in identifying southern European opportunities, however (*The Economist*, 'Spanish pickings', 9 July 1989).

Defence

Although not 'officially' on the 1992 single-market agenda, and clearly more a potential than actual candidate for the dropping of national purchasing preferences, European defence and especially defence electronics firms appear to have motivated to prepare for a more market-integrated day. For, if pressures from '1992' *per se*—or the vision of an eventual European defence community—are not (yet) strong, limits on national procurement budgets and the widespread perceptions of a (much) reduced Soviet threat are (see *The Economist*, 'Fighting Allies: Electrical Brief', 4 March 1989.)

One result has been a proliferation of collaboration agreements and minority shareholdings across borders such as those between British Aerospace and France's Thomson, Plessey's 49 per cent of Italy's Elettronica and GEC (United Kingdom)'s 5 per cent in France's Matra. The largest 're-alignment', however, has perhaps been that of Daimler–Benz's agglomeration of AEG and (pending) acquisition of MBB, thus creating a single large national defence supplier in West Germany.

Few 'new moves' from America firms appear much in evidence in Europe in defence or aerospace. A long-rumoured link between McDonnell Douglas and the Airbus consortium for collaboration on a four-engined, long-distance plane to compete with world-leader Boeing has not come to pass. However, General Electric, with its long-standing joint venture with France's SNECMA for jet-engine production, appears well situated to take advantage of any European aerospace successes. And Northrop's minority position in a newly dynamic Fokker of the Netherlands many provide that American firm with some counterbalances to United States' defence cuts.

Food processing

The food industry, thanks to the European Court decisions concerning Germany's barriers to imports of *crème de cassis* and beer was the birthplace of the principle of 'mutual recognition' of national standards. Business strategies here, too, have hardly been unaffected. Once again, the most striking single move, the acquisition by France's BSN of the vast bulk of Nabisco's European operations, was made possible by the sell-off of European activities by an American firm. As a result of the leveraged buy-out of RJR–Nabisco, the new shareholders, KKR, sold off European subsidiaries in order to raise funds to pay back and service the debt with which the leveraged buy-out was made (*Financial Times*, 'Nabisco to sell five divisions', 10 May 1989, p. 27). The result is the domination of Europe's food industry by a 'big three': Unilever, Nestlé (which made its own highly publicized acquisition of Britain's Rowntree in the late 1980s) and BSN.

Pharmaceuticals

With many 'NTB' (Non-tariff barrier) subsidiaries spotted around a Europe of national insurance and purchasing authorities, and with national health regulations compounding the fragmentation, the pharmaceutical industry has at least the potential for significant '1992' rationalization. Notwithstanding, the allowance by the European Court of 'health and safety' standards as an escape clause to the principle of 'mutual recognition', as well as a pharmaceutical industry well habituated to a NTB-ridden market—and thence differential pricing—suggests that movement will be slow. Still, there has been one major industry re-alignment already, yet again involving an American firm. The 'merger' between Beecham's of the United Kingdom, newly slimmed down of its consumer products' activities, and SmithKline-Beckman of the United States is in its essentials arguably a strategic move by Beechams to obtain (a heretofore lacking) sales and distribution position on the Continent via the take-over of a temporarily troubled American pharmaceutical multinational already having such a network!

Motor vehicles

Both the genesis of 'Project 1992' itself, and the competitive dynamics of the motor industry in Europe are inseparably linked to the threat of Japanese penetration which has stalked the European industry since the early 1970s. 'Restructuring, rationalization and consolidation' moves by European and American firms in the EC have long been taking place in response to Japanese thrusts. Japanese imports first were aimed at the 'non-auto-producing', smaller countries of Europe, such as Belgium, the Netherlands, Ireland, Denmark, Greece and Switzerland, where Japanese market shares are now in excess of 20 per cent. Then came assembly ventures such as those of Nissan for small trucks in Spain, and Honda's far-reaching 'cooperation' accords with the United Kingdom's Rover Group. The global dimension of the competitive battle is underscored by recalling that one of the major European 'restructurings', the sale by Chrysler of its European marketing and production network to Peugeot of France in the late 1970s, was the direct result of Chrysler's near bankruptcy in the United States—inflicted at the hand of the Japanese. European fears of the Japanese were further reinforced by European defeats in the lower-priced, volume-car segments of the America market for imported cars. By the mid-1980s Volkswagen's position in the United States had become a pale shadow of the glory days of the Beetle, and Renault was deep in the ignominious process that later led it to withdraw from the American auto market by ceding control of American Motors to a revitalized Chrysler in 1987.

Already set in motion in reaction to the severe recession of the early 1980s, European corporate moves toward 1992 have been a mix of cost-cutting retrenchment and automation, continued trade protection of the Italian, French, Spanish and British markets from Japanese imports, geographical re-alignment of production locations and innovative model development and marketing—especially by 'specialist' producers such as BMW who attempt to 'stay above' the volume-car fray.

Volkswagen—the number one volume leader in Europe, although hotly pursued by Fiat and Peugeot—appears set to become the leading 'Euro-producer', increasingly moving production to Spain, all the while cutting costs and automating in Germany. Interestingly, much of the management undertaking this Euro-rationalization process has been brought to VW from Ford Europe—the first auto producer to 'treat Europe like one big plant'. Fiat, Peugeot and Renault have pursued essentially 'nationally based', 'automation first (plus continued trade protection from the Japanese menace)' production strategies in Italy and France—albeit from lower labour—cost bases than VW's Wolfsburg home. (Limitations of Japanese imports into Italy and France one recalls, are severe—1 per cent market share in Italy and 3 per cent in France. The former in particular are an ironic result of negotiated *mutual* quotas by Japan and Italy dating from the 1950s and Japanese fears of small Italian cars crushing nascent Japanese production . . .!) Peugeot, Fiat and Renault have, however, just lost their fierce battle to prevent the EC Commission from abolishing these national quotas on Japanese imports as of 1992 (*Le marché financier de Paris*, 9 June 1989, p. 5). Will these firms' response in turn be 'European', or will they turn to ever more global sourcing of components and parts in the NICs of the Third World in order to remain cost-competitive with the Japanese?—thus creating yet another global spillover effect of 'Europe 1992?'

As hinted above, an American firm, Ford, was the 'original' Euro-producer. Indeed, perhaps the very first '1992' moves in the industry were the large Spanish plants of Ford and General Motors, clearly intended from the first to be exporters to the EC, and opened in 1977 and 1983 (*The Financial Times*, 'Motor Industry Supplement: Spain', 20 October 1988, p. IV). Notwithstanding such American company strengths in Europe, the '1992 strategies' of Ford and GM appear to be of the 'hold and maintain market share (at about 12 per cent and 10 per cent, respectively) variety'. The American market—reinforced by the American 'profits-now (or very soon) mentality'—maintains pride of place in these American firms' thinking. In particular, General Motors' failure to gain a share in Europe, in spite of instigating a bout of price-cutting competition in the mid-1980s, as well as its continuing loss of market share in its home market, suggests few assertive moves in the near future.

Japanese firms continue, however, to knock at, or rather leap over, the gate: Honda through its Rover collaboration, plus plans to export soon-to-be-on-line American production to the EC; and Nissan and Toyota from new 200,000 per year, full-scale assembly plants sited in the United Kingdom (*The Financial Times*, 'The Toyota effect . . .', 19 April 1989 and 'Honda takes fresh aim at Europe', 9 March 1989). There appears no willingness here to 'trade off' tacit 'spheres of influence', with the Japanese staying in Asia in exchange for Europeans staying home. Of course, the European market for autos has now well surpassed that of what used to be the world's largest market, the United States. The potential for continued Japanese challenges to the European motor industry, and, what is more, the generation of ever-renewed trade-policy conflicts between the United States' and EC trade authorities, as well as between the United Kingdom and the southern Europeans, is thus more than manifest.

The debate over what amount of 'local content' does or does not make a product 'European' is unlikely to go away soon under such circumstances.

Microcircuits and computers

Among all the industrial sectors over which European government officials, and many business leaders, display technological and competitive *Angst*, the microcircuit and computer industries are at the top of the list. The historical model of European response first to American, then threatened Japanese domination of these industries was—in the larger countries at least—the very archetype of the 'national protection and national subsidies to nurture national champions' policy. That policy manifestly failed. American firms still dominate European markets for both computers and microcircuits, primarily through European-based production, and the Japanese dominate world markets for semi-conductor 'chips'. The 'European company' share of Europe's chip market—itself but 18 per cent of world semiconductor demand—is estimated at some 38 per cent. (*The Economist*, 18 February 1989, p.74). In computers, the 'European' share of the European market may be as much as 50 per cent (*Financial Times*, 26 May 1989, p. 18), but two of the firms once seen as Europe's best hopes, Nixdorf and Norsk Data, have fallen on hard times, and Britain's ICL, Germany's Siemens and France's Bull depend on procurement preferences and R & D subsidies to maintain much of that share. Moreover, in spite of the global competitiveness problems of American firms in other industries, here the locus of innovation, be it in microcircuits, workstations, PC's, software, networking or super-computers, clearly remains primarily in the United States—with Japan rising strongly.

Part of the reason that the old 'national champion' model failed was that the European firms were obsessed by computer mainframe and basic memory chips during the 1970s, to the point that they competed head-on with the then leaders, IBM in mainframe computers and Texas Instruments and the Japanese 'big fives' in memory chips. The European 'champions' hardly anticipated, nor were their politically influenced and structured firms able to foment, the radical alteration and fragmentation of the computer industry by first the mini, then the PC, then the networking and now the workstation revolution. Perhaps because of a preponderance of national military customers, and a consumer electronic market less dynamic than that of Japan, the European semi-conductor firms also did not flourish—or lead the way into microprocessors—in a nationally protected and subsidized environment.

Much of the European business thrust for Project 1992 has roots here, for Philips is Europe's leading semi-conductor firm—and was one of the partners in the failed 1970s' consortium of national champions in computers called Unidata. Opening the market aims not only at scale and scope, but at stimulating—or eliminating—sleepy national champion firms and developing European global competitors.

It is here, however, where the sceptics, and American and Japanese and other Asian outsiders, concerns become manifest, for there is another possible outcome, to wit, that national protection and subsidization of coddled champions becomes metamorphosed into 'Fortress Europe' plus subsidization of 'regional champions'. Two of the major events on the road to 1992 are major projects to subsidize European chipmaking: 'Megaproject', with a $2 billion price tag for taxpayers, started in 1985 to aid Philips and Siemens, and 'Jessi' (Joint European Semi-conductor Silicon Initiative), with a price tag of $4.5 billion (half of which is to come from the companies) over eight years (*The Economist*, 18 February 1989, p.74; *The Financial Times*, 21 June 1989, leader page).

While American and Japanese 'outsiders' fulminate and strategize against 'Fortress Europe', insiders also raise doubts. A *Financial Times* editorial notes:

Similar 'strategic industry' arguments have been used in Europe in the past to justify government support for a variety of industries. Their record has been dismal. What reason is there to think that Jessi, half of which is due to be financed by the European Commission and national governments, will be any more successful? [21 June 1989]

Ultimately, competition down on the corporate playing field will determine the outcome, and the Europeans would be very foolish to imagine that a European 'Fortress' can be constructed around what has already become an international industrial crossroads, inextricably linked to the United States and Japan—even were trade occasionally to be impeded by complex, strict or mercurial 'rules of origin'. The Americans are already 'in' via local production, or, like Intel, shortly will be. So, for that matter, are several of the Japanese, with NEC and Fujitsu with major facilities in the United Kingdom and Ireland—and major expansions announced—as well as ownership (NEC-Bull) and technological links (Fujitsu-ICL) with European firms (*The Financial Times*, 13 April 1989, p. 20). In and around the increasingly active, even frenzied flirting for partners, positions and acquisitions, lurks the possibility of takeovers—even by the dreaded Japanese (*The Financial Times*, 'Japanese may soon begin friendly takeovers in the EC', 5 May 1989). And the Americans may have a few surprises up their sleeves in what would be the shape of the world semi-conductor/computer industry if IBM were to start marketing its microcircuits to outside customers? What will be the eventual impact of the PC/ workstation/networking revolution? And will computers, chips and the like become commodities anyway, with the 'value added' (why else care about 'hightech?') increasingly in software, not hardware? Or will American firms prove complacent about Europe, or simply ignore it as they have in other industries in the past, thus providing opportunities in the future for European advances?

In European companies' headquarters, intra-firm flirtations and political lobbying appear currently to predominate (*The Financial Times*, 'A high-tech high noon approaches', 26 May 1989, p. 18.). The most significant corporate move in computers has been made by France's Bull, one which gives the firm a supranational character—across the Atlantic. It has taken control of Honeywell US and, with a minority stake in Japan's NEC, is trying to establish a 'co-headquarters'—Paris – Boston Route 128 response to the global competitive challenge. Probably the most significant move in semi-conductors has neither been the combination of SGS (Italy) and Thomson (France), nor Jessi nor Megaproject, but Philips's not very public strategic integration into its global strategy (directed from the Netherlands) of its long quasi-independent American activities, which include Signetics, one of the leading semi-conductor producers in America.

Conclusions

From a kaleidoscope of facts, patterns emerge. What can be said about European company strategies? First, there are a sufficient number of examples of European firms actually responding to the challenge of Project 1992 to suggest that the 'environment will not be re-negotiated' by purely defensive mergers, alliances and joint ventures—or by the re-imposition of trade barriers by other names. Second, European firms now have something that American firms woefully lack, a galvanizing vision of how to improve their international competitiveness in a world of *global* competition. Moreover, it is a vision that was developed and communicated by European business leaders themselves, not by politicians and generals. Third, a surprising, even astounding number '1992 related moves' have already spilled over onto the American competitive playing field—global competition is the name of the game now being played, not some vague slogan to which lip service, but not much else is paid.

The Japanese? Perhaps somewhat belatedly, distracted as they were by the conquest of American markets, they now seem to be aware that Europe 1992 has meaning for the game of world market share via cost and quality leadership they have been playing so skilfully for so long. They hardly seem so exhausted by their American victories, however, that they will have won their 'battle of Stamford Bridge' only to be defeated shortly thereafter in European combat. If the price of staying in the European market is 'to become good corporate citizens' (translation: comply with written or unwritten local or EC content rules), they are showing that that price will be paid. How successful the Japanese will be in European competition will largely depend on how well, and how rapidly, they can in fact make the hardly insignificant shift from strategies based on exporting to Europe to producing in it. It will also depend on the degree to which the Japanese do not confuse Europe 1992 and a single European market with a homogenous

European market in which nuances of taste and custom, as well as all creative means to maintain some national preferences by other names, will also have disappeared. And it will also depend on how long they can continue to use British or Irish production sites as a trampoline for access into the rest of the EC in the face of annoyance or opposition from other member countries, or on whether they can spread out production to other countries in the Community. A 'single market' is most unlikely to mean a 'single polity' and a single political reaction to inward foreign investment in the foreseeable future. And the French are smarting from their defeat by the Commission over their wish to replace national quotas on Japanese car imports with EC quotas.

American corporate strategic response to 1992 is as heterogeneous as the continental United States itself. Taken as a whole, it is, however, none too reassuring. The very heterogeneity highlights a weakness: American business has no galvanizing mission for global competition—not '1992', or 'export or die', or 'dominate world markets', or any other—let alone a mission designed and articulated by business itself. Indeed, many of the moves toward consolidation—or advancing—of market share positions in Europe by European firms have been facilitated by American firms, retreats from Europe. And the reasons for these retreats—low near-term returns in telecommunications, need to pay off junk-bond debt from a takeover, financial problems at home due to Third World debt problems for banks etc.—have an aura of temporal, geographical and strategic myopia about them. Even in sectors of traditional American international strength, such as semi-conductors and computers, one wonders whether American firms have thought sufficiently about the ramifications of the '1992 moves' in the global competitive game which are showing up, or will show up, on American shores.

American firms have yet another handicap in formulating strategy for 1992. The reshuffling of industry in Europe is very likely to provoke a game of gaining market share for long-term, low-cost production advantage. This is not the same environment as that into which the 'American Challengers' of the 1960s settled in—and from which they profited relatively soon after establishing 'reasonable' stakes in national markets. If the single market is to mean anything, 'me-too' players with 'reasonable' stakes will get shaken out, and the dominant positions will eventually accrue to those who are patient enough, and prepared enough to accept low margins for a time, to achieve those dominant positions.

There are some American companies with positive strategies for Europe 1992, but are there enough to slow or halt the slide of American firms global competitive position which has been going on since the 1960s? (see Franko, 1989). American business vision spanning the Atlantic seems in short supply. Can it be mobilized at all? The success of American firms in Europe—and in United States itself—depends on it.

References

The Economist, 'The Gyllenhammar Group', 4 July 1987, p. 67.
—'A Survey of Europe's Internal Market', 9 July 1988.
—'Spanish pickings', 9 July 1988, p. 72.
Franko, Lawrence G., *The European Multinationals*, London, Harper and Row, Stamford, Conn., Greylock Press, 1976.
—*The Threat of Japanese Multinationals: How the West Can Respond*, Chichester and New York, Wiley, 1983.
—'Global Corporate Competition: Who's Winning, Who's Losing, and the R&D Factor as One Reason Why', *Strategic Management Journal*, Fall 1989.
Jeelof, Gerrit, 'Importance of the 1992 Program of the Information Technologies Sector', speech given at the Conference Agenda 1992, The Fletcher School, Tufts University, 7 April 1989.
New York Times, 'Affluent German Unions Fear 1992' (8 March 1989, p. D1.
d'Oultremont, Patrice, 'Comments on Information Technology: Impact of the Single Market', Conference on 'Agenda 1992', The Fletcher School, Tufts University 7 April 1989.

11 Signposts on the road to trade policy reform in agriculture
C. Ford Runge

The complex nature of agricultural protectionism has long served as a barrier to popular and political understanding—a state of affairs useful to the special interests that benefit the most from this protection. The inconsistencies, inefficiencies and inequities that riddle farm policies have usually been of interest only to a relatively small group of specialists. As President Kennedy reputedly told his principal agricultural policy adviser: 'I don't want to hear about agriculture from anyone but you . . . . Come to think of it, I don't want to hear about it from you either.'

The currently unsettled world agricultural trade situation may be leading to chaos or to a new order. The outcome of the United States current disputes over hormones with the European Community and in the General Agreement on Tariffs and Trade (GATT) may be to liberalize trade or to plunge world markets into a cycle of protectionist retaliation and counter retaliation. The parties to the GATT have begun to debate the farm trade issue as part of a broad effort to reform the global trading system. As the GATT discussions of farm trade accelerate, it is clear that agriculture must now take a central place in United States foreign trade policy; reform in agricultural policies can be the stimulus for broader trade liberalization and reform. If farm policy is to be liberal, rather than protectionist, it must reflect greater market orientation both at home and abroad, and reduce the distortions that now separate the domestic from the global markets. Renewing growth through trade, important to the industrialized economies, is even more important to developing countries, where agriculture remains the foundation of economic progress.

Success in this effort will require broader understanding of the domestic causes of agricultural trade distortion. It will require the political courage to discard some of the myths that have protected ineffective and obsolete policies. It will require coordinated, multilateral efforts to rationalize domestic agricultural policies through international agreements. Otherwise, agricultural trade disputes may lead to increased global protectionism and stagnating growth.[1]

There are three important signposts in this process. Together, they will determine whether world agricultural trade reform moves forward or takes a turn toward greater protectionism. The first is labelled 'Geneva, April 1989', where trade ministers gathered to try to salvage agreement from the failures over agriculture at the Montreal summit in December, 1988. The second is labelled 'Uruguay Round/Farm Bill Debate, 1990' and involves the interaction of the concluding year of the GATT round and the 1990 Farm Bill debate scheduled for the same period. The final signpost, somewhat distant, but still clearly visible, is labelled 'Europe, 1992'. Let me consider each signpost in turn, then, looking backward from 1992, attempt to draw some insights for the likely course of events.

The road not taken in Montreal

The Montreal trade talks ended in December 1988 as they began in 1986; with the United States and the EC squarely opposed on agriculture, both threatening to up the ante in the agricultural trade conflict. In the end, there was essential agreement in twelve of fifteen negotiating areas, including trade in services and various measures strengthening the GATT system. Trade ministers agreed to meet again in April 1989 to try to salvage the three main areas that remained in disarray: textiles, intellectual property rights and agriculture. These disagreements, especially in agriculture, threatened all other progress, since the rule of the summit was 'nothing is agreed until everything is agreed'.

The other areas of conflict were thus linked, directly and indirectly, to agriculture. Textiles is a game the United States plays not to win, so much as to stay even. Trade in textiles is governed by a complicated compact called the Multi-Fibre Agreement (MFA) which protects America manufacturers through quotas on imported clothing. Contrary to its calls for more liberal trade in agriculture, American textile policy has been to protect the status quo.

Trade in intellectual property rights (TRIPS in GATT jargon) involves patents and other business in information and ideas. Like textiles, it pits developed against developing countries. Unlike textiles, which is a declining industry in most developed economies, TRIPS tends to involve high-technology industries like communications and computers. If a California company makes and copyrights a video cassette, for example, this copyright may not protect it if the law does not apply in other countries. Yet many LDCs argue that in the early stages of American and European development, new technologies were often 'pirated', reputedly including the porcelain rollers first brought from Hungary to grind Minnesota wheat into fine breadmaking flour.

In the late stages of the Montreal talks, India and Brazil demanded concessions on TRIPS that were unacceptable to the United States and the EC. At the same time, the United States and the EC were both playing for the support of the developing countries on agriculture. Because GATT decisions are made by consensus, any country can halt the play on any item. Demands by India and Brazil on intellectual property were thus linked to textiles—one of their most important export sectors. Nowhere is this kind of linkage more obvious than in agriculture, which emerged in Montreal as the hub around which all other negotiating games revolved.

Geneva, April 1989

With failure to agree on agriculture in Montreal, the April meeting in Geneva was the first signpost to watch. The Bush administration's new trade representative, Carla Hills, stepped into a complex and demanding situation that immediately tested both her toughness and knowledge of agricultural trade issues. She and Clayton Yeutter at Agriculture faced experienced, resolute European negotiators who have been at the game of agricultural trade talks for decades. Fortunately, the negotiations were very ably handled by Warren Lavorel, of the United States Trade Representative's Office and were aided greatly by the behind the scenes' work of the GATT Secretariat, and its head, Arthur Dunkel.

The key features of the agriculture agreement reached in Geneva were:

(1)   A promise to make 'substantial and progressive reductions' in farm subsidies.
(2)   No increase in overall farm subsidies in 1989.
(3)   No increase in restrictions on agricultural imports in 1989.
(4)   By October 1989, GATT nations are to reveal plans for cuts in 1990 farm supports.

Influencing the April discussion was the hormones dispute. On 1 January, 1989, the EC announced a ban on all beef imports from the United States containing hormones used to help fatten cattle. Citing health risks, the EC action touched off a cycle of retaliation. This apparently isolated example of health regulations acting as trade barriers is part of an emerging pattern of environmental and health issues with major consequences for the world economy. These consequences are especially important to trade between developed and developing nations, and will also figure prominently in the 1992 debate, I will mention.

Uruguay round/farm bill, 1990

Encouragingly, April brought accommodation and willingness to move forward in the GATT. The stage is now set for progress on domestic policy reform during 1989–90 in Washington. But a reverse dynamic—toward less liberal trade and greater protectionism—is also possible if progress in the GATT stalls. Keep in mind that the most protected sectors in agriculture would find a trade war very helpful in the short run. Elected representatives of these special interests would benefit from trade conflict by emphasizing the short-term rewards of protection and their own 'toughness'.

In general, direct payments to farmers will remain the principal alternative to current policies, although movement in this direction will be difficult. The key negotiating objective should be the phased elimination of those measures with the greatest distorting effects on trade, and the substitution of an income 'safety net' in their place. It is unnecessary to maintain that the final outcome be no supports whatever, if the principle of phased elimination to much lower levels of price support can be established as a binding obligation under the GATT. The overall effect, especially if acreage retirement is also eliminated except on fragile lands, will be to enhance diversified cropping mixes, creating demand for research into new crop alternatives and widening the base of agricultural productivity gains.

A second guiding principle is that social welfare objectives of agricultural policies should be a part of the discussion in the GATT. These may appear tangential, but can be crucial to selling policy reforms to domestic interests and relating them to various titles of the 1990 Farm Bill. In the United States the agriculture committees of Congress must be able to convince their constituents (both commodity and consumer groups) that the United States has received a 'fair deal' in the GATT. This deal must find acceptance with farmers and the non-farm public alike.

Selling trade reform can only be done as part of a package, the bulk of which involves domestic policy issues. In my view, domestic agricultural policy must be expanded to encompass rural employment issues and enviromental quality objectives, together with farm income security, as the three pillars of policy. Trade reform then fits into this package of benefits. Linking trade policy reforms to environmental and rural employment objectives in the 1990 Farm Bill can not only enhance the attractiveness of the overall package, but its impact on dollars flowing into rural development and revitalization.

Third, the developing countries must be convinced that agricultural trade reforms are in their interest. Despite the advantages of liberalization, many of these Third World producers remain skeptical. Insofar as possible, these governments can be encouraged by the promise of market access. A real opportunity exists to bring the developing countries into the same framework of increased market orientation as the developed countries (albeit in different ways), especially if market access is tied to interest write-downs on debt. The role of the GATT in improving the developing countries' market mechanisms may be even more important to growth in world trade than reforms in the agricultural economies of the developed world. Removing government disincentives to production designed to favour urban elites, especially in Africa, would be worth billions of dollars in foreign development aid.

Given the entrenched interests and the profound difficulties facing agricultural policies, however, the temptation to blame foreign governments for failed domestic policies remains strong. Negotiators in both the United States and the European Community can engage in recriminatory changes and counter-charges, followed by renewed subsidy wars, without serious short-term losses in domestic political terms. The trade situation can thus lock into a domestic political environment in which high levels of trade conflict make continued price supports and policy distortions at home appear politically desirable.

Europe, 1992

The last and most distant signpost is labelled 'Europe, 1992'. A recently published study for the National Planning Association [2] notes both opportunities and threats in this development. In brief, the EC is attempting to remove the barriers among its twelve member–states to the free movement of goods, services, capital, and people—a process called 'completing the internal market'. Despite a slow start, momentum was achieved by early 1988 and an irreversible process begun under the so-called Single European Act. The programme is contained in a Brussels White Paper, which lists about 300 measures or areas requiring action. The barriers targeted for elimination fall into the following categories: border controls; restrictions on the recognition of professional qualifications; differences in value-added and excise taxes; legal regimes; restrictions on the movement of capital; restrictions on services; regulations and technical standards; and public procurement markets. The formal adoption in 1985 of a comprehensive programme, which included a timetable for action on specific measures, faces an overall deadline of 31 December 1992.

It is noteworthy that the only sector in which this process is already largely complete is in agriculture, under the Common Agricultural Policy (CAP). If the CAP is a harbinger of 1992, protectionism will increase, not decrease, as member countries in the EC trade-off barriers between them for barriers *vis-à-vis* the rest of the world.

A dynamic trend in the CAP has been a drift into protectionism, as harmony among EC member–states is purchased at the price of non-tariff barriers applied to outside competitors. [3] Europe is increasingly confident of her ability to trade *en bloc*, and many feel it is willing to be regarded as collectively protectionist in order to capture the large benefits of an internal market.

In my view, one of the most important realms of trade conflict between the United States and the EC was signalled by the hormones dispute. In the high-income countries of Europe and North America, health, safety and environmental regulations are especially attractive candidates for use as non-tariff barriers.

They are part of a larger problem: enviromental and health risks are increasingly traded among nations along with goods and services. While increasing emphasis is given to the growing role of services' trade, comparatively little has been directed to risks which are the opposite of services: environmental and health *disservices* traded across national borders. This problem arises directly from the transfer of technology, and will increasingly affect international investment flows, product liability, trade and development and the relative competitiveness of American business.

The particular temptation to convert health, safety and environmental regulations into protectionist trade barriers is not confined to the EC. Similar developments have been noted in the United States, part of a growing division in the regulatory structures of developed and developing countries.

Consider these developments:

- During 1988, the American Soybean Association conducted a campaign emphasizing the health risks of palm oil, which competes directly with soybean oil in the processed food market. Palm oil is produced entirely in developing countries, notably Malaysia and coastal Africa. Malaysia has sought to enjoin the soybean growers from conducting the campaign.
- In February 1989, the Natural Resources Defence Council released a report citing significant health risks from the use of Alar, a growth retardant, on America apples. While United States' regulatory agencies may ultimately ban its use, no controls are in force over continued applications outside the United States. The Alar episode coincided with a scare over Chilean fruit (a major source of American winter supplies) that resulted in a temporary ban on these fruit imports.

The emergence of a two-tiered international structure of environmental regulation results from increasingly stringent rules and regulations and a rising concern with environmental quality and human health among wealthy nations. In most developing countries, however, rapid economic growth remains the primary focus of concern. This creates incentives to export restricted industrial materials—or whole production processes—from North to South. A kind of 'environmental arbitrage' results, in which profits are gained by exploiting the differential in regulations. In the United States, for example, the Federal Insecticide, Fungicide and Rodenticide Act (FIFRA), the Safe Drinking Water Act (SDWA) and the 1990 Farm Bill are all likely to be amended in ways that effectively constrain chemical and land use choices. These are but several examples which may lead multinational firms to expand in markets where regulatory oversight is less constraining.

This environmental arbitrage results from conscious policy choices that reveal differences in the value attached to environmental quality by rich and poor countries. As these paths of institutional innovation increasingly diverge, so will the differential impact of environmental constraints on business in Europe, North America and, say, Argentina and Brazil. The competitiveness implications of these trends are not lost on developed country firms. They have been quick to see the trade relevance of environmental and health standards in limiting access both to developing country competition and other developed countries. Growing consumer concerns with health and the environment create a natural (and much larger) constituency for non-tariff barriers to trade. It is doubtful, for example, that beef-offal merchants in the European Community could have blocked competitive American imports solely in the name of superior French or German beef kidneys. But the hormones question created a large, vocal and committed constituency for denying American access to this market. These distortions threaten more liberal international trade in ways that are damaging to both developed and developing country interests, yet are not widely appreciated in the economic and legal communities.

Beyond agriculture and health and safety regulations, what other prospects does 1992 offer? According to the NPA study, they are mixed.

- While some reduction in border controls will continue, problems relating to different taxes, plant and animal health requirements, and control of drug trafficking, terrorists and immigration will prevent complete elimination of the controls.
- Agreement was recently reached for mutual recognition among the member states of professional qualifications; but there is likely to be continued resistance.
- Differences in indirect taxes will be particularly difficult to achieve because they directly involve member government's revenue and entail surrender of control over taxation to Brussels.
- Action is likely to be completed on a Community trademark, is somewhat less likely on the Community patent and a start will be made on copyrights.
- Although the EC has agreed to phase out all remaining restrictions on capital movements by 1992, in view of the threats to the weaker currencies (for example the lira) presented by complete liberalization, it is likely that there will be some backsliding.
- Major changes are expected in financial services, although progress is expected to be slower in deregulating transportation and the advanced communication technologies.
- Although a major attack is being mounted on restrictions to public procurement, opposition will remain strong and limit effective implementation.

Despite this somewhat mixed review, the potential for change in the EC resulting from completion of the internal market is considerable. An EC commissioned study estimates potential benefits as great as a 7 per cent increase in gross domestic product, a 6 per cent reduction in prices and the creation of five million jobs.

The American economic stake is also considerable: the EC accounts for 25 per cent of American exports and 40 per cent of its foreign investment, which contributes impressively to Europe's economic activity. The creation of a single EC market presents America business with potential for increased demand, lower costs and potential economies of scale. However, these opportunities will also be available to EC and third-country firms.

It is essential that the United States' government closely monitor developments in the EC to avoid actions that would adversely affect American interests. Problems are bound to arise.

In responding to these problems, the negotiating approach of the Bush administration will largely follow that of the Reagan years, although it may shift its emphasis somewhat from Europe to Japan and the Far East, consistent with the background and experience of its new negotiators.

Independently of American demands for reform, the completion of the internal market will have pervasive effects on European agriculture.[4] Harmonization of policies in the area of indirect taxes may reduce the costs of agricultural supplies, although this effect will vary across member–states. The transportation industry, as it becomes more competitive, will decrease the cost of food and feedstuff transport. The single-market principle will also be inconsistent with many 'side-payments' to member-states made under the CAP, including the complex 'monetary compensatory amounts' (MCAs) used to adjust the impact of the CAP on different national currencies. As integration proceeds, the CAP will become less the focal point of a unified Europe, and its special status as a symbol of unity will decline, perhaps aiding the process of reform. Finally, and perhaps most importantly, higher growth rates in the EC manufacturing sector would allow excess labour in agriculture to be more easily absorbed in other sectors.

Looking backward, 1992, 1990, 1989.

Looking backward from 1992, the intermediate prospects for the GATT and domestic agricultural policy reform look difficult, but not impossible. It is likely that European farm lobbies will continue to resist reductions in overall levels of subsidies, although budget pressures in the Community will continue to force agricultural belt-tightening. Looking across the Atlantic at a newly unified North American trading bloc, Europe sees serious competition requiring comparable measures. I remain optimistic that, in spite of these difficulties, it will be possible to clear a path to trade policy reform. But it cannot be done easily, nor, as I hope I have emphasized, without foresight.

References

1. C. Ford Runge, 'The Assault on Agricultural Protectionism.' *Foreign Affairs*, Vol. 67, No. 1 (Fall, 1988), pp. 133–150.
2. Michael Calingaert, 'The 1992 Challenge from Europe: Development of the European Community's Internal Market', National Planning Association No. 237, Washington D.C. 1988.
3. C. Ford Runge and Harald von Witzke, 'Institutional Change in the Common Agricultural Policy of the European Community', *American Journal of Agriculture Economics* 692, 1987.
4. Y. Leon and L. P. Mahé, 'The CAP after 1992: A Fairly Common Agricultural Policy', paper presented at a symposium on 'Europe 1992: CAP Reform and World Agricultural Trade', American Agricultural Economics Association, Annual Meeting, Baton Rouge, Louisiana, 1 August 1989.

12 Will Japan seek regionalism?

Susumu Yamakage

Introduction

Seen from the very Far East, the European Community seemed just a fragile lid covering traditionally antagonistic nations that were very different from one another with respect to language, culture and history. There were trade disputes between Japan and the EC like the export restraint of steel made in Japan, but they were subdued, at least in Tokyo, by such bilateral conflicts as France versus Japan on VCRs and Perrier or Britain versus Japan on Scotch whisky etc. in terms of the salience of negotiation. The EC as an economic entity, as distinct from a mere aggregate of West European countries, used to be a real concern only to few specialists and a handful of Japanese multinational corporations (MNCs) operating in the region.

For quite understandable reasons, attention of the Japanese government with respect to foreign economic policy has been directed mainly toward Washington, DC. Furthermore, Tokyo seems to prefer to deal with the sovereign states within such frameworks as G-5, G-7 in addition to the G-2 or United States–Japan coordination. Tokyo's relationship with the EC is sometimes a complicating, rather than simplifying, factor in its policy *vis-à-vis* West European countries.[1]

The unification plan of the twelve national economies of the EC had interested the Japanese little until very recently, and hence its vast and profound consequences were understudied, if done at all. Response to the Single European Act was slow. This contrasted greatly with the fact that the EC's economic unification was stimulated partly by the threatening competiveness of Japanese industry and technology. Since late 1988, however, manufacturers, investors and bankers as well as journalists have been increasingly concerned with the prospects of the EC after 1992. It is nearly impossible to fail to find at least one article on Japanese interests in the EC in each issue of economic magazines or daily newspapers.[2]

Today, Japanese economic transactions with the EC region are roughly a half of those with the United States in terms of exports, imports, Japan's direct investment in the respected areas and vice versa. But, Japan's interest in the EC's future is higher than such figures indicate. Notwithstanding the lack of a clear image of the post-92 EC, Japanese companies are now studying the basis of their operations in the region, and the Japanese government does not hesitate to show misgivings concerning potential protectionism or isolationism.

The EC's preparations for a unified market are imposing a new complication in Tokyo's concern with Japan's prospects in the world economy. The movement of the EC toward 1992 is influencing the Japanese perception of the structure of world economy more significantly and profoundly, if indirectly, than recurrent trade friction between the EC or its members and Japan. In other words, the post-92 EC is considered not as an isolated independent movement in Western Europe, but as one of several related movements that reflect the current trend toward regionalism. Under these circumstances perceived, or real, the Japanese are increasingly preoccupied with their own version of regionalism in order to respond to such a trend.

This short chapter will discuss the relationship between the EC and Japan from the above perspective. Namely, instead of focusing on direct interaction between the two, this chapter will try to elucidate the meaning of the scheduled further integration of the EC for Japan's future. Specifically, it will discuss Japan's perception of the changing world economy's trend toward regional compartmentalization and its attempts to cope with such a trend. It will argue three points: (1) the international economic environment surrounding Japan is unmistakably tending toward regionalism; (2) though Tokyo's ostensible outlook has been globalism, it attempted, and failed, to take initiatives toward regionalism; and (3) Japan's regionalistic options are now wider, but the choice may be difficult. In this perspective, the EC affects Japan's future position in various ways.

The tide of regionalism around Japan

In the last few years, important developments which may affect Japan's future economic outlook have been taking place one after another. Some are global, some regional; because of their global importance, much attention is paid to the former. Therefore, a short remark will suffice here.

The EC's movement toward 1992.

The consensus among the leadership of the EC countries to unify their economies was undoubtedly one of the most historic decisions in the post-war international political arena. This unification will realize an ideal which was proclaimed three decades ago, but which at the time was hardly believed to be likely at least in this century. It still remains to be seen if unification will be completed by the end of 1992, especially until the EC summit in June 1989 is assessed. Even if unification is delayed as most of the observers predict, the direction will not be reversed. The impact upon Japan will be multi-fold. As sketched at the outset, Japanese industries will be variously affected by the unified market and by the unified policy. Japan's major trading partners, the United States and Asian capitalist economies, are linked with the EC more closely than Japan is. The impact on them will eventually affect the Japanese economy.

United States–Canada integration by 1999

While it is a bilateral arrangement,the agreement to form a free trade area between the United States and Canada will create a North American free trade area by the twenty-first century. This move by the United States is much more significant than a similar one with Israel some years ago. The United States–Canada agreement is considered an unmistakable signal of American's departure from globalism toward regionalism at least in relative terms.[3]

In the immediate neighbourhood, too, the Japanese are observing several new, and potentially crucial, phenomena that could affect their regional concerns. Some notable ones are as follows.

Economic growth of Asian NIEs and some ASEAN members

The economic performance in the 1970s of Asian Newly Industrializing Economies (NIEs) comprising South Korea, Taiwan, Hongkong and Singapore was so impressive that those economies were compared to uprising dragons. While their economic growth remained high throughout the 1980s, some of the member countries of the Association of South-East Asian Nations (ASEAN) have followed the same path.[4] Thailand and Malaysia in particular are experiencing rapid industrialization and economic growth. These economies along the Western rim of the Pacific used to depend heavily on the American market, but a genuine economic interdependence among themselves is now taking shape.[5] The appreciation of the Japanese yen and the rapid restructuring of Japanese industries accelerated Japan's direct investment in the region by the manufacturing industry, and hence transfer of technology and production capacities toward it. Japanese imports of semi-finished and finished products from the NIEs–ASEAN region are steadily increasing with only a slight time lag. Backed by a huge trade surplus, the investment by Asian NIEs' corporations in the ASEAN region is increasing in intensity.

The American approach to NIEs and ASEAN

Under the above mentioned circumstances, the United States' government recently abolished the Generalized Scheme of Preference (GSP) treatment *vis-à-vis* Asian NIEs and will reportedly reduce the preference regarding some manufactured goods produced in Thailand in the near future. Washington also started to discuss the possibility of a free-trade agreement with ASEAN, although a speedy development in this area is not anticipated at the present time. Due to heavy dependence upon the American market, NIEs and ASEAN countries have to respond to American initiatives.[6] On the one hand, they may accept concessions in favour of the United States in the areas of intellectual property, service trade, voluntary export restraint, foreign exchange rate etc.; on the other hand, they are exploring the possibilities of some form of regional institution that would confront potential American protectionism and secure their access to the American market.

China toward 1997

The leadership of the People's Republic of China (PRC) invented the 'one state, two systems' concept so as to solve the dispute with Britain over the status of Hong Kong. Hong Kong, one of the Asian NIEs, is to come under the sovereignty of the PRC in 1997. Because some areas already in China, especially Shanghai and its vicinity, are practically close to the NIE status in terms of economic infrastructure and outputs, the political unification of the PRC with Hong Kong may well legitimize China's fitness to be called an NIE. The viability of the regime of 'one state, two systems' or its impact upon the surrounding region and upon China itself is, needless to say, yet to be seen. Nevertheless, the Beijing government seems to be contemplating unification with Taiwan, another Asian NIE, with a similar scheme in mind. No matter if the current economic reform results in successful development or social and political instability, China will become a pivot of economic restructuring in East Asia.

China, the Soviet Union and the Pacific Community

Although the global economic focus is moving toward the Pacific, there is no comparable equivalence to the Atlantic community there. The most comprehensive, but still very nebulous, framework is a semi-official institution named the Pacific Economic Cooperation Committee (PECC).[7] While its basis is economic, consisting of such market economies surrounding the Pacific as the United States, Canada, Japan, South Korea, Taiwan, the ASEAN countries, Australia etc., the PECC conveyed the political implication that it symbolized economic consolidation among non-communist, if not anti-communist, countries many of which belonged to the Western Alliance. However, drastic economic reforms at home and normalized diplomatic relations with the United States made it possible for the PRC to become a member of the PECC. Now, claiming to be a Pacific state, Moscow has shown a keen, and unusually positive, interest in membership in this loose institution. Although this raised controversies among PECC members who more or less represent official views of their home-governments, a *rapprochement* between Beijing and Moscow might pave the way for the latter to join the Committee. Should the Soviet Union become a member of the PECC, the framework of eonomic cooperation in the Pacific would be characterized as trans-ideological interdependence. The trend is obvious: the cleavage of the post-war Cold War has become almost, if not completely, obsolete in the area of economic interdependence. Despite the lack of diplomatic ties with either the PRC or the Soviet Union, for instance, anti-communist South Korea started direct economic dealings with these communist giants.

In short, global and regional trends surrounding Japan are leading toward regionalism. As a matter of fact, the world has already long been in an age of regionalism. Since the late 1950s, many international organizations have been established (some of which have already disappeared from the scene, however) in Asia, Africa, Latin America and Oceania, not to mention Europe, for the purpose of some sort of regional economic integration. Today, there are some twenty regional organizations for economic cooperation/integration to which more than 120 countries, out of the total of 170 in the world, belong in some way or another. Among major powers, China, Japan and the United States were the only exceptions—that is until recently. With the United States becoming a part of a North American free-trade area, East Asia remains the only region that fails to have any regional economic integration scheme. Will the Japanese seek some scheme? Will they take the initiative?

Japan's past initiatives and lessons

The Japanese have been widely accused of having a narrow, egocentric economic mind and a reluctance to taking initiative and responsibility for the international order. These criticisms are not entirely groundless. However, if they imply that the Japanese have neither had any will nor attempted to take initiatives toward the creation of regional arrangements, they are based on distorted images and stereotypes.

It is true that as a nation that committed war crimes during World War II, the Japanese were long reluctant to take any initiatives abroad, and Japan's post-war involvement in Asia began with its war reparations. When Japan was admitted to the club of advanced countries in the mid-1960s, however, it started to exercise its limited power, that is economic power, in the creation of a stable and prosperous Asia, and environment favourable to the Japanese. It is of utmost importance to remember Japan's attempts and their consequences in order to understand Japan's future roles, either expected abroad or self-appointed, in the international arena.[8]

Partial failure in the ADB

The first significant attempt to influence development of regional institutions has been visible since the early 1960s. Namely, Japan took the lead in the formation of the Asian Development Bank (ADB) in the United Nations Economic Committee on Asia and the Far East (ECAFE, now ESCAP). Providing a contribution equal to that of the United States, the other largest donor, Japan expected other Asian nations would welcome the leadership of the only industrialized nation in Asia. It turned out, however, that they expressed their resentment of Japan's aims by choosing Manila, instead of Tokyo, as the bank's headquarters, which the Japanese had not anticipated at all. Tokyo's proposal for an increased contribution to the ADB, which would exceed the American contribution, and hence voting power, was refused, and a special fund which did not affect the power structure in the board was set up with Tokyo's donation alone. Tokyo succeeded in sending a Japanese to be the first president of the ADB, and so far only Japanese have occupied the presidency of the ADB, but the Japanese learned that their influence was not yet welcomed.[9]

Patronage of the MCEDSEA

Shocked by an embarrassing defeat, the Japanese government was quick to aim at improving Asian sentiments towards Japan, and to institutionalize Japan's influence in Asia in a less controversial way. Accordingly, Tokyo convened the Ministerial Conference on Economic Development in Southeast Asia (MCEDSEA) in 1966 or barely a half year after the set-back on the ADB. The conference decided to reconvene once a year thereafter hosted by a different participating government on each occasion. Tokyo tried to utilize the MCEDSEA to buy South-east Asian governments' general support of Japan in exchange for the latter's official development assistance (ODA). Tokyo was careful to obtain American approval of Japan's initiative, and Washington welcomed Japan's burden sharing,: that is, in defence of non-communist Asia, Japan would exercise its economic power to support development policies of South-east Asian countries while the United States would be able to concentrate more on a military–strategic commitment in Asia. Despite American support, South-east Asian leaders disliked policy dialogue with Japan at the MCEDSEA. The majority of the members preferred to utilize their own collaborative institution, ASEAN, in order to deal with Japan collectively. Since 1974 the conference has not been convened, and Tokyo allowed it to be forgotten. The exercise of power through a positive sanction of ODA, which Tokyo thought was less controversial, also turned out to be unpopular among Asian neighbours too.

Pacific cooperation initiative

Subsequently, Japan found another less controversial means of influence. The key concept was interdependence, which signified two-way influence as opposed to the one-way influence of foreign aid. The target area was Asia-Pacific, which was much broader than South-east Asia and included the United States so as to dilute Japan's influence. It seemed a very promising institutional framework, at last to the Japanese, and they sought an official start in the late 1970s. However, misgivings, if not resentments, arose from some corners in Asia. The idea was abandoned. Subsequently Tokyo proposed a more decentralized, loosely structured, less formal institution representing both official and private sectors. At last, the PECC was established along this line. Here, government officials are supposed to represent not the government but themselves. As mentioned earlier, the organization was unmistakably pro-Western, but it succeeded in attracting not only the PRC but also the Soviet Union. On the other hand, Japan's influence was diluted as expected.

Since then, Japan has not attempted to take an official initiative in forming a regional institution. The government had taken initiatives, and in general had received negative responses. Tokyo learned to be passive and to adopt a low-posture approach in proposing international schemes involving Japan's power in some way or another.

Without doubt, Japan increasingly affects the regional economy in Asia-Pacific. Tokyo tripled its ODA during the last decade, and Asia enjoyed the lion's share. Tokyo's plan to recycle Japan's huge surplus funds globally will undoubtedly affect Asia. Nevertheless, Tokyo still seems hesitant to launch any idea of either a comprehensive or a functional institution for regional economic cooperation/integration. As Japanese interests have spread to various regions in the world, Tokyo has emphasized globalism.

Japan's options

Throughout the 1980s, Japan's options with respect to regionalism have been quite limited. In fact loosely structured Pacific economic cooperation has been the only realistic option open to Japan, and the Japanese have accepted this. Recently, however, some Japanese have begun to perceive a wider range of options. Such views are increasingly shared among economists, industrialists and bureaucrats. Regionalist approaches are now discussed openly. However, the substance of regionalism remains unclear. Although the Asia-Pacific region is usually referred to as a possible framework for Japan's regionalism, it is unclear if Asia-Pacific includes, for example, the United States or China, not to mention the Soviet Union or Latin America. The expected role of Japan is unclear, too. The concept is so nebulous as to require a careful examination of definition and context.

Tokyo's external economic policy has not changed at the fundamental level: the pursuit of globalism is Japan's foremost goal. On the other hand, a review of the current outlook has been undertaken beneath the surface. Some detailed differences aside, there are at least four major options that are conceivable to the Japanese.

Toward a Pacific community

This is in a sense an extension of Tokyo's current policy, and seems the most acceptable option as a future choice. The Chinese and the Russians are positively interested in Pacific economic cooperation, and this makes the scheme much less vulnerable in the political arena. Most, if not all, of the major powers in the Asia-Pacific region would participate in the scheme, and would consequently make it so multilateral that smaller countries will be less threatened by the potential domination of a single major power. The circumstances are more favourable than ever for Tokyo to suggest upgrading the semi-official PECC to a genuine inter-governmental institution for Pacific cooperation. There is a hint of disillusionment among the Japanese participants in the PECC, however. [10] The organization is already multilateral enough to make it extremely difficult to reach agreement on every major issue. The Japanese are not listened to as much as they had hoped. Thus, the road towards the Pacific community is yet to be well paved.

United States–Japan alignment

An alternative to slow-moving Pacific cooperation is a bilateral, presumably quick-moving, alignment with the United States. Encouraged by American advocates including notably the former Ambassador to Japan, and fearful of American protectionism, some Japanese are now seriously considering the possibility of some comprehensive agreement with Washington on free trade between the two countries. The major issue for the Japanese is how to cope with American pressure based on the principle of reciprocity. In fact they are concerned less with the formation of a free-trade area across the Pacific than with the establishment of a permanent organ to deal systematically with trade conflicts that never seem to subside.[11] As for Japan, however, it is difficult to create a more or less exclusive arrangement with the United States at the expense of its Asian partners. As for Washington, a trade agreement with Japan is not the only option; the Americans are exploring several bilateral arrangements with Asian countries besides Japan. The situation appears ironic because these two largest economies are strong advocates of a global approach to a more liberalized international economy, at least at the governmental level.[12]

Leadership in the western Pacific rim.

A third alternative is for the Japanese to emphasize and strengthen their ties with their neighbourhood.[13] Unlike the 1960s or 1970s, the relationships between Japan and other Asian countries are more or less equal now, and will be more so in the decade to come. Economic interdependence between Japan, Asian NIEs and ASEAN countries is rapidly deepening in terms of direct investment, manufactured commodity trades, and so-called division of labour. Theoretical reasoning aside, some Japanese see an intertwined development pattern in those economies that will eventuate in the creation of a network of interdependence in the western Pacific rim in the near future. The importance of the American market will consequently be diminished, as the advocates of this policy see it, to the extent that it is no longer vital to the Japanese economy. In their eyes, the region is increasingly viable in terms of substantial economic integration, and a formal institution to promote such a movement in the private sector is useful and desirable.

Japan's economic hegemony

The last option is for Japan to take over American responsibility for the management of the international economy at least in Asia, probably in larger areas in the world than Asia. Actually, this option is suggested much more by foreigners than by the Japanese themselves. As the major creditor and the major ODA donor, Japan could more than ever exercise its economic power, certainly *vis-à-vis* developing countries in Asia-Pacific and probably in North and South America. Also, Japan's position *vis-à-vis* other aid-giving countries, notably the United States, would be strengthened. Tokyo would then be much more responsible for the stability and prosperity of economies in the region, including the American economy. While the Japanese themselves do not believe this is a feasible future option, it is usually mentioned in the context of a warning about Japan's alleged ambition.

The above four options, some of which are more feasible than others, can be summarized in terms of (1) which area Japan's regionalism would cover, and (2) in a region defined as such, what kind of leadership Tokyo would assume and to what extent. As for (1) a bilateral arrangement with the United States covers perhaps the smallest geographical area while economic activities to be covered would be extraordinarily large. The last option would cover the largest area. As for (2) the last two options would require Tokyo to exercise more power than the first two.

Constraints to Japan's Pursuit for Regionalism

Japan's options toward regionalistic institutionalization have certainly widened, but are still limited. The following constraints especially seem to limit Tokyo's choice considerably.

Heavy dependence on the American market

No matter what kind of regionalism is sought by the Japanese, their dependence on the American market must be one of the most crucial constraints. Their exports to the United States amounted to $83 billion or 36 per cent of total exports in 1987. Reduced access to the American market would impose a devastating impact upon Japanese export industries. Diversifying their exports from America to smaller markets would be difficult not only because of the huge amount and volume of exports, but also because a single product, the automobile, occupies a significantly large share. In this respect the EC constitutes a second large market, but it is unlikely to replace the United States.

Japan's slow opening of its market

For Asian NIEs and ASEAN countries, the Japanese market is as important as America's if not more. Not only Americans but also Asians are calling strongly for much freer access to the Japanese market. Realizing that the liberalization of external transactions is not sufficient to open the market, the Japanese recently committed themselves to taking further necessary measures. They seem to acknowledge not only that the current situation is unfair to the foreigners, but also that opening up the Japanese market is beneficial to themselves too. Agricultural products including rice are no exception. Both public opinion and the government are now less sympathetic to excessively protected Japanese farmers.[14] The Japanese increasingly regard the opening of the market for agricultural and manufactured products as in their national interest. Unfortunately, the pace of abolition of domestic barriers and obstacles will not be so fast as to satisfy foreign exporters. It seems, therefore, unlikely that Japan will succeed the United States as Asia's export market in the near future. In other words, Asian countries will remain dependent not only on the Japanese but also the American and EC markets.

Improbable yen bloc

The formation of an economic bloc without the United States may be closely associated with the formation of the Japanese yen bloc in the western Pacific rim. Although the Japanese yen is far behind the dollar, the internationalization of the yen is steadily progressing.

The Japanese yen is being used increasingly in the region for trade, investment and other forms of economic transaction, pushing up the demand for the yen by governments, corporations and individuals throughout the region. If the yen plays a more important role outside Japan, however, the use of both the yen and the dollar becomes much more realistic. Some Japanese advocate a dual currency system of settlement in the region should a formal institution be formed under Japan's initiative. It is highly unlikely that Asians will accept a yen bloc at the expense of delinking with the United States' dollar.

Refusal of Japan's initiative

People in Asia still remember the Japanese invasion of their homelands, and few seem to forgive Japan's wrongdoing even after nearly a half century. Japan is not fully trusted, and often is unintentionally and unexpectedly the target of criticisms, some of which are harsh, some others just subtle. Today's Japanese are partly responsible. Their insensitive and inconsiderate words and deeds have jarred on neighbours' nerves and prevented the decay of old memories. Whenever Tokyo proposes a regional framework involving Japan and its neighbours, the government has to be very careful not to recall the nightmare of the East-Asian Co-Prosperity Sphere. Among others, the PRC, South Korea and some ASEAN countries tend to eschew Japanese leadership in the region. In addition, they are very well aware that to remind forgetful Japanese of past militarism enhances their bargaining position *vis-à-vis* Tokyo.

Other calls for Pacific cooperation

Recently, other Asian countries besides Japan have become active in launching ideas for Pacific cooperation.[15] South Korea will soon establish formal ties with ASEAN as the United States, Japan, Australia and a few other powers have already done. Moreover, Seoul proposed its own version of an Asian-Pacific cooperation agreement. Some ASEAN leaders are suggesting collective action by consolidated Asia-Pacific countries *vis-à-vis* the EC and the United States, and they were probably inspired by ASEAN's success in dealing with powers outside the region. Recently, Australia proposed her version of an economic organization. Even Taiwan is said to be interested in the creation of an East-Asian common market. Such moves indicate increasing interests among Asian countries in the formalization of some sort of regional scheme. This may or may not turn out to be favourable for Japan's initiative. While a proposal from Tokyo, would gain positive attention of the leadership of nations in the region, it may well conflict with other proposals based on different interests or outlooks.

Advocacy of globalism among the Japanese

Finally, the Japanese would prefer a global regime in the international economy which they hope would prevent the proliferation of protectionism. There is a strong reluctance to take the lead in regionalism as a matter of Japanese self-interest. As briefly sketched above, the situation is not favourable to the Japanese. Nevertheless, Tokyo wishes to take part in the Uruguay Round of the GATT, and use its influence to stop further regionalization. Even Japan's various types of foreign aid, which used to be heavily allocated to Asia, are being spent increasingly in other regions. Not only Japanese commercial activities but also Japan's official commitments are becoming global. The Japanese have their own barrier to the pursuit of regionalism: the dilemma of globalism or regionalism.

None of those constraints sketched above seem likely to disappear in the near future. Some are built into the Japanese economy and its structural interdependence with foreign countries; some are based on external political, economic, psychological and emotional factors. The Japanese have to allay foreign concerns at least to a certain extent, whatever option they may choose for Japan.

Globalism versus regionalism

The Japanese, both government officials and businessmen, are seeking a way toward not only (1) securing the global institutionalization of a liberal economic system and free trade, which is now in trouble; but also (2) facilitating the settlement of international economic conflicts so as to ease tensions between Japan and its major partners. Tokyo is aware that these tensions endanger the liberal economic system. In search of those two goals simultaneously, Tokyo is actively involved in existing global institutions such as the IMF, the GATT, etc.

The conflicting issues in the GATT Uruguay Round are not comparable with those in the preceding rounds. While the past rounds focused primarily on manufactures, the current round involves agricultural trade, service industries and non-tariff barriers such as intellectual property and industrial standards. It will take more time to overcome differences and conflicts of interests, if it is successful at all. A perception of urgency is growing among the Japanese. The mounting problems with their major trading partners are now seen as too critical to wait for a global agreement. Consequently, Japan seems forced to explore some kind of regional arrangement in a more formal way.

Both Japanese government and business are seriously concerned with a growing protectionism in America and Europe and anxious to prevent the emergence of economic blocs. Japan will try to avoid any action that might animate regionalist tendencies further. To date, there is no concrete plan of regionalism on the Japanese side. It seems safe to say that the Japanese still prefer a global approach to secure a stable and prosperous world economy. Nonetheless, they are seeking some sort of regionalism that will be compatible with globalism.

However, although Japan now has a wider range of options, none of these seems preferable to the others to an obvious extent. Having learned lessons from past experiences of attempted leadership, the Japanese government is still reluctant to speak out on its own initiative. Current constraints to Tokyo's pursuit of any one of the options are too complex to suggest any promising option. For the time being, Tokyo has to attempt the careful melding of Japanese interests with those of others in order to gain support for whatever scheme of regionalism is in the offing.

Given the above mentioned options and constraints as well as Japan's increasing commitment to international responsibility, there *is* a feasible scenario that Japan might profitably try to pursue. Should it need to seek regionalism, the Japanese government could pursue a three-level goal with respect to building up some sort of regional arrangement surrounding Japan. These levels are mutually complementary, rather than exclusive. If all were accomplished, the Asia-Pacific would become a huge area of institutionalized interdependence. While Tokyo is not optimistic, it could try to pursue them all simultaneously.

As a primary goal, Tokyo would seek a formal arrangement with the United States which must be a more or less comprehensive agreement. The American market is indispensable, and close ties with the United States are the best security for Japan. At the same time, however, Tokyo must not let the scheme be protectionist *vis-à-vis* Asian exporters, since this would not only evoke harsh criticism but also go against Japanese interests in Asia.

Second, Tokyo would attempt to enhance cooperation in the Western Pacific rim as a source of pressure *vis-à-vis* the United States and the EC in order to counter their respective protectionist movements. This would also serve the enhancement of intra-regional economic interdependence, with further liberalization of the Japanese market. Japan would make every effort to join such a cooperative framework in order to prevent collective action toward Tokyo. At the same time the Japanese government would avoid strong leadership in order not to invite resentment against the nation.

Third, Tokyo would also attempt to promote bilateral and other arrangements between the United States and Asian countries so as to build up the set of bilateral agreements into a network across the Pacific. If a comprehensive multilateral agreement is difficult to achieve, bilateral networks become an alternative approach. Stated simply, Asia-Pacific interdependence has become trilateral with the United States, Japan and the western Pacific NIEs-ASEAN region as the three corners. It would be in Tokyo's interest to institutionalize the trilateral interdependence that already exists.

The EC in the context of Japan's response to regionalism

Faced with pressing international economic problems, the Japanese are concerned especially with two considerations: the first relatively short-term and the second relatively long-term and presumably more serious. The first concerns the increasing economic disputes with major trading partners, including the EC as an entity as well as some of its member countries; the second is missing the train named 'regionalism' which the EC has been pulling and on which most people are now riding or about to ride. The EC's movement toward 1992 is needless to say related more to the latter concern. In so far as Japan's own regionalist interests are concerned, the EC will remain secondary. Tokyo must pay more attention to Washington, Beijing, Bangkok, Jakarta and probably Moscow than to Brussels, London or Paris. Nevertheless, the EC will occupy a certain, not insignificant position in relation to Japan's response to the changing global economy.

While the Japanese have begun to explore the possibility of regionalism involving Japan, their economic activities are already too widespread all over the world to be contained in the Asia-Pacific area. Even if Japan joins a regional economic integration/cooperation area, it will remain a global economic power. The EC cannot be ignored when thinking about Japan's future position in international society and of the trade prospects of Asia-Pacific countries.[16]

The Japanese are seriously worried about the increasingly influential trend toward protectionist regionalism. In this regard, the EC is a main source of such a threat, for Japan's trade with the EC countries will suffer from protectionistic import restrictions. Between 1983 and 1987 both exports and imports with the EC region doubled in value consequently doubling Japan's trade surplus from $10 to $20 billion. Japanese trade may indeed nurture protectionism for manufactured goods within the EC. Moreover, Tokyo fears that the EC may take the lead in the proliferation of protectionistic regionalism. In this regard, the EC's denial that it will become 'Fortress Europe' has been undermined by its own adoption of 'balance of mutual benefits and reciprocity'. The rise of protectionism in one EC country may trigger similar movements in other EC countries, and might eventually induce the EC to take unified protectionist action against Japan.

As they did to counter the trade imbalance with the United States, Japanese manufacturing companies have begun to transfer production capacity into the EC region. While Japanese direct investment in the region used to be mainly in the field of finance and insurance, it is now prominent in manufacturing. A unified market of 320 million people with a $4 trillion GNP is certainly attractive. Japanese MNCs' increasing investment can be also explained as a pre-emptive move before the EC becomes a fortress.

The EC ought to be positively affected directly as well as indirectly by Tokyo's more or less independent efforts to prevent the proliferation of protectionism. Partly in response to American and other external pressures and partly because of Japan's own determination, Tokyo will open the Japanese market in order that foreigners may have easier access to manufacturing and service industries as well as direct investment. Both farmers and manufacturers in the EC region will benefit. EC-based MNCs will benefit too.

The EC could encourage Japan to play a global role rather than to force it to opt for regionalism. For the EC, and for the United States too, Japan's commitment to global responsibility must be desirable in many respects. Japan's ODA, especially grant-in-aid, is increasingly allocated to Africa where EC countries have been deeply involved. Also, African countries should be beneficiaries of Japan's surplus re-cycling plan. Tokyo would be willing to contribute to the Lomé system. The EC could utilize Japan's resources, that is money and technology, for the purpose of the economic development of the ACP countries.

In conclusion, it may be appropriate to recall that the EC originated from the endeavour to secure perpetual peace in Europe. It proved in the age of nation-states the possibility of the 'community' overlying different nations. The global inclination toward regionalism that we are witnessing today is largely promoted by the attempt to institutionalize deepening economic interdependence. But, the sense of community is an essential condition for successful regionalism. The Japanese have failed to recognize fully the crucial importance of an inter- or trans-national community. Economic influence is not enough to obtain membership. The Japanese are being asked if they can exercise their power as a member of an international community, either global or regional.

References

1. For the background of Japan–EC relations, see Loukas Tsoukalis and Maureen White, *Japan and Western Europe: Conflict and Cooperation*, New York, St. Martin's Press, 1982; and Masamichi Hanabusa, *Trade Problems between Japan and Western Europe*, New York, Praeger, 1979. Both are a little out-dated, but provide a good summary. The latter author is a Japanese diplomat. Also see Albrecht Rothacher, *Economic Diplomacy between the European Community and Japan, 1959–1981*, Aldershot, Gower, 1983.
2. For instance, in March 1989, a Japanese weekly magazine, *Ekonomisuto* (Economist) published a 170-page special supplement entirely devoted to issues on the EC after 1992, entitled 'Oshu no Chosen: 92nen EC Togo to Nihon Keizai' (the Challenge of Europe: the 1992 Integration and the Japanese Economy). In the March/April 1989 issue, the *Journal of Japanese Trade and Industry*, a magazine which more or less, but not necessarily, reflects the view of the Japanese Ministry of International Trade and Industry (MITI), chose 'Europe After 1992' as a cover story.
3. Jeffrey J. Schott (ed.), *Free Trade Areas and U.S. Trade Policy*, Washington, D.C., Institute for International Economies, 1989.
4. Colin I. Bradford Jr and William H. Branson (eds), *Trade and Structural Change in Pacific Asia* (NBER Conference Report), Chicago, University of Chicago Press, 1987.
5. Current changes in economic relations among Japan, Asian NIEs and ASEAN countries are so rapid and dynamic that the brief description here is based not on solid analysis but on information obtained from various newspaper reports and papers submitted to conferences the author attended in recent months.
6. Schott, op. cit.
7. The origin and nature of the PECC is explained in the section entitled 'Pacific cooperation initiative' below.
8. The argument in this section is partly based on Susumu Yamakage, 'Asia-Taiheiyo to Nihon' (Asia-Pacific and Japan) in Akio Watanabe (ed.), *Sengo Nihon no Taigai Seisaku* (Post-War Japan's Foreign Policy), Tokyo, Yuhikaku, 1985.
9. For an extensive study of Japan and the ADB, see Dennis T. Yasutomo, *Japan and the Asian Development Bank*, New York, Praeger, 1983.
10. This author's personal interview with some Japanese members of the PECC conducted in April 1988.
11. Keen interest in the United States–Canada Agreement expressed in Japan is directed toward the establishment of the panel to settle economic disputes, rather than free trade in itself.
12. See Makoto Kuroda's contribution and comments to it in Schott, op. cit. He is MITI's former vice-minister for international affairs.

13. Rokuro Tsuchiya (ed.), *Nihon Keizai no Kokusai-ka to Asia Keizai* (the Japanese Economy's Internationalization and Asian Economies), Tokyo, Yuhikaku, 1987; Toshio Watanabe, *Nishi-Taiheiyo no Jidai* (An Era of Western Pacific), bungei-Shunju-sha 1989.
14. Although strong sceptical comments are expressed about Japan's opening market by some Japanese as well as many foreigners, the Japanese public are increasingly aware of the necessity, and even desirability, of such actions. See an official public opinion poll on the economic structural adjustment published in *Gekkan Seron Chosa* (Monthly Public Opinion Survey), March 1989.
15. This paragraph is based on various reports carried in Japanese newspapers between late 1988 and early 1989.
16. *Far Eastern Economic Review*, 18 May 1989, pp. 68–73 which focuses on the impact of the EC after 1992 upon Asian economies.

Part V The EC and the Third World

13 1992 and ACP trade prospects*
Carol Cosgrove

1992 and the ACP: An Introductory Overview

The formation of the Single European Market and 1992 are hailed in Europe as a window of opportunity for the developing countries. Africa, Asia and Latin America, however, watch and wait with some trepidation. There is uncertainty as to how to attempt to assess the likely implications of 1992 on their trade and investment policies regarding the European Community, the most important market for their products. They fear that the Single European Market will constitute a threat rather than an opportunity. There is a widespread belief that the member–states of the EC might well become more inward-looking in seeking to promote access to opportunities within each other's economies, at the same time endeavouring to protect their markets from potential threats from other neighbours.

Many developing countries fear that their traditional trade relations with individual EC member–states might be harmed by the movement towards 1992 without being replaced by equivalent special relations with the EC as a whole. In their view the danger is that new, and for them not easily surmountable, non-tariff barriers could be set up around the EC. Their fear is that as the weakest trading partners, they have most to lose.

For their part, spokesmen for the EC persistently point to the opportunities which 1992 and a Europe without internal frontiers will offer in terms of new markets for developing country exports. The Cecchini Report on 'the cost of non-Europe' estimated that the European economy would be greatly stimulated by the process of forming the Single European Market and that in theory, consumers would look outside for more products. European Commission spokesmen also tend to emphasize the advantages which an integrated market will bring in terms of greater transparency of trading conditions due to the harmonization of health and safety regulations as well as tax laws. The contention is that Europe's trading partners should be able to produce greater quantities of exports at lower cost.

Most third world countries, however, are somewhat sceptical. In their view,

*The author would like to thank Paul Campayne of the University of Reading Graduate School of European Studies and Laurie Walter of Tufts University (London campus) for undertaking much of the research for this chapter. In addition, Dr George Yannopoulos and Irene Malloy of CTA Economic & Export Analysts Limited have provided invaluable assistance.

the boom in European trade will mainly benefit the Europeans themselves. They believe that their experiences with the Common Agricultural Policy (CAP) do not provide much room for optimism regarding the actual organization of the Single European Market. The fear is that as more and more industries in Europe become 'sensitive', powerful and vocal lobbies in the member–states may become more aggressive in their search for special safeguards.

The African, Caribbean and Pacific (ACP) states constitute a special group in terms of their trade relations with the EC. Since 1975 they have benefited from the Lomé Convention, a trade and aid agreement between the EC countries collectively and the now sixty-six ACP states. The ACP are some of the least competitive of Third World exporters, and may stand to lose a great deal from 1992. They export primarily raw materials and unprocessed commodities to the EC, products with low-price elasticity of demand. The implication is that growing incomes within the Single European Market will probably not noticably increase demands for ACP exports. The ACP states, moreover, are unlikely to be able to turn readily to other markets due to the uncompetitiveness and inflexibility of their exports. Many of them depend upon almost a monoculture for survival, eight commodities accounting for over 80 per cent of their earnings.[1]

The British Minister of Overseas Development, Christopher Patten, has called the Lomé Convention 'the most liberal trading regime on offer to any group of developing countries by any industrialised partner'.[2] Through the Lomé Convention, ACP states enjoy duty and quota-free access to EC markets for all their industrialized products except rum. All their agricultural products not covered by the CAP enjoy duty-and quota-free entry and there are special arrangements for bananas, beef and sugar.

The year 1992 is regarded with especial foreboding by those ACP states exporting sugar, rum and bananas—all three commodities being subject to national quotas in the EC market. Some of these products provide individual ACP states with almost 90 per cent of their foreign exchange earnings. The EC's sugar import regime is probably safe for the foreseeable future, but the situation is less clear regarding bananas and rum.

The present banana regime is not compatible with the objectives of the Single European Market. It is based on different national quotas which exclude the free circulation of bananas. France ensures access only to bananas from former French colonies (mainly The Ivory Coast and Cameroon) and its *départements d'outre mer*. Italy protects its banana imports from Somalia. Other member states, especially Greece, protect their own production. Germany on the other hand has a duty-free quota which covers its total consumption, principally from Latin America. The United Kingdom protects its market against Central American bananas: the Commonwealth Caribbean countries are permitted to export all their bananas into the United Kingdom duty-free, while Central American producers are subject to import quota. Currently, Commonwealth

Caribbean suppliers account for 17 per cent of the British Market. The British government has committed itself to maintaining the present regime. It is seeking safeguards from other Community countries to ensure that after 1992 Commonwealth Caribbean bananas will continue to enter the Single European Market in their present volume.[3] Yet a solution has still to be found. In particular, West Germany is not agreeable to a substantial increase in its traditional prices; a view with which the Netherlands concurs.

The Caribbean banana producers themselves realize that they can no longer rely on preferential access for their traditional exports. The Prime Minister of Dominica, Miss Eugenia Charles, stressed the need to improve efficiency and quality to ensure that Commonwealth bananas could begin to compete effectively with their Central American counterparts. Bananas play such a significant role in the small island economies, however, that it is hard to see how any free circulation of bananas in a Single European Market could not have marginal yet disastrous consequencies for economies such as those in the Windward Islands. In many instances, moreover, it is these same economies which would be most adversely affected by any dramatic changes in the European Community import regime for rum, which is yet to be the subject of negotiation.

The EC insists that ACP states will be more than compensated for any losses stemming from 1992 in terms of the stimulation which the Single European Market would exert on their exports of value-added products. Indeed, the EC has emphasized that a successor to the present Lomé Convention should move its emphasis from agricultural and traditional aspects of economic development towards more diversified value-added non-traditional sectors. Recent studies undertaken at Reading University, however, indicate that there is little evidence to date to suggest that the Lomé Convention has had a beneficial impact on ACP non-traditional exports for the EC market. The remainder of this chapter examines the impact of the Lomé trade regime on ACP value-added exports and considers the question, 'Does Lomé Work?'

The Pre-Lomé regime

The Lomé Convention is the centrepiece of the EC's policies towards developing countries. The origins of the Convention lie in Europe's colonial legacies. When the European Economic Community (EEC) was founded, France and Belgium had extensive colonial possessions in Africa and elsewhere. The establishment of the EEC, with its Common External Tariff, meant that they could no longer maintain preferential trade relations with these 'Overseas Countries and Territories (OCTs). The French therefore applied intense pressure to find an accommodation in the new EEC for continuing economic relationships with her OCT's. This led to the establishment of a temporary, five-year, 'association' between the original six EEC states and their dependent OCTs, in which preferential access for OCT exports to all EEC member–states was given. Before this association expired, the new independent African countries pressed the EEC to extend the special relationship. The EEC therefore negotiated a new agreement, the Yaounde Convention, signed in 1963, which continued the membership and main provisions of the previous association. This was in turn replaced by Yaounde II, which ran along similar lines until 1975.

The accession of the United Kingdom in 1973 to the EEC was the next major step in the development of the association agreements. Britain insisted on special arrangements for the developing countries in the Commonwealth, who had until then benefited from tariff-free exports to the United Kingdom under the system of Commonwealth preference. The extension of the association to all these countries was ruled out on the grounds that it would have so diluted the Yaounde system as to make it of little value. Instead, all those states whose economic structure and production were deemed by the EEC to be comparable to the existing associates were offered the chance to participate in the successor to Yaounde II. The countries excluded were the Asian Commonwealth (Bangladesh, India, Singapore, Pakistan, Malaysia, Sri Lanka, and Hong Kong) who instead signed cooperation agreements with the EEC.

The remaining ACP Commonwealth countries, together with the existing Yaounde Convention countries and some other African states, negotiated a new association agreement with the EEC. This first Lomé Convention, signed in February 1975, was in operation from 1976 to 1980. Primarily to coordinate their activities *vis-à-vis* the EEC, a formal ACP group was created in June 1975. A second Lomé Convention was negotiated for the period 1980–5, and a third for 1986–90. The latest Convention associates an EC of now twelve member–states with sixty-six independent ACP countries. Negotiations are presently underway for a fourth Lomé Convention, and for the eventual accession of an independent Namibia.

The Lomé trade regime

The basic Lomé trade regime provides for duty-free access to the EC of nearly all products originating in the ACP states.[4] Moreover, no reciprocal tariff preferences are demanded from ACP countries. The exceptions to this duty-free access are ACP exports of CAP products. These are still subject to levies, but the reduced rate represents better treatment than that afforded to almost all other countries. Sugar, rum and beef are covered by special quota arrangements.

A key aspect of the regime is that the products must originate in the ACP states. Goods wholly manufactured in the ACP states clearly originate from them. Products using materials imported from third countries present a definitional problem to which 'rules of origin' apply. ACP producers have to meet 'process criteria', requiring the substantial transformation of inputs from one tariff heading to another for the final product. In addition, they must meet further process and/or value-added requirements if the products come under 'list A' attached to the rules of origin of the Lomé Convention. Tariff-free treatment is therefore given only where typically 50–60 per cent of the value of the final product is added locally. Some ACP producers see the rules of origin as a non-tariff barrier. This is especially the case for list A, products deemed sensitive by the EC, such as textiles, footwear and wood products. All are products where ACP countries may possess a competitive advantage over EEC producers, and therefore could be used to diversify exports away from primary products. However, the rules of origin are probably less restrictive than those other developing countries face in Western markets. Moreover, components and material from the EC or from other ACP countries, also count as local.

The benefits of the Lomé Convention have to be seen in the wider context of other international trade arrangements.

International trade arrangements

General Agreement on Tariffs & Trade (GATT)

Most international trade takes place under the international regulatory framework of the GATT. Its specific role is to facilitate arrangements for the substantial reduction of both tariff and non-tariff barriers to trade. The two basic principles governing the operation of the GATT are non-discrimination and reciprocity. These are subsumed within the most-favoured-nation clause. Important exceptions to these principles were made for customs unions and free-trade areas, and special derogations for international trading arrangements such as the Lomé regime. Since 1979, however, trade preferences for developing countries have been formally allowed under the GATT, without the need for non-discrimination and reciprocity.

The seven rounds of multi-lateral trade negotiations under the GATT have led to substantial reductions in world tariff levels. The present Uruguay Round seeks further trade liberalization, tariff reductions, removal of non-tariff barriers and its extension into agriculture and services.

The General System of Preferences (GSP)

The GSP was set up by UNCTAD in 1968. It accords preferential tariff treatment to the exports of developing countries in the markets of the industrialized countries without requiring them to make counter concessions. The GSP schemes of different countries/markets differ widely in terms of content, scope and applications, but there are a number of common features such as tariff reductions, safeguard clauses and rules of origin. Moreover, quotas and ceilings on the amount of imports permitted under the schemes have been set particularly in areas where the manufacturing exports from developing countries have the greatest potential, for example textiles.

In the case of the EC, various rates and quotas apply to agricultural imports. Industrial products are rated zero or low. Sensitive items are subject to individual country ceilings and/or quotas. The EC scheme seeks to maintain benefits for developing countries that rely on GSP access to the EC to ensure that competition does not harm domestic producers and to 'keep faith' with ACP states and the Lomé agreements. The ACP states interpret this to mean that their preference levels should be protected. Thus in their view, the GSP should not be extended.

Multi Fibre Arrangement (MFA)

The MFA seeks to manage trade in the textile and clothing sectors. Under it, importing countries negotiate restrictions on imports of specific items, and unilaterally place quotas on products if certain conditions are met. In 1982, the EC states set an overall quota for textile and clothing imports from ACP states, known as the 'Ligne ACP'. After intense negotiations it was abolished in January 1987. Nevertheless, any significant increase in the level of ACP exports could lead to pressure for so-called 'voluntary export restraints'.

Impact on Lomé

Although nominally generous, it can be seen that the trade concessions and access to the EEC market enjoyed by the ACP states are not as substantial as may appear at first sight. The combined effect of the GATT, GSP and the CET means that in any event over three-quarters of ACP exports to the EC would meet zero tariffs. This is because the great bulk of ACP exports are tropical foodstuffs and raw materials. In the case of industrial products, more than 90 per cent of ACP exports would be eligible for GSP anyway. The one advantage of Lomé compared with GSP in this respect is that there are no formal quantitative restrictions. Where the EC does impose tariffs on products coming from third countries of export interest to the ACP, the general level of those tariffs is low and may be further reduced by the Uruguay round.

The question therefore arises whether the specific trade provisions of Lomé in fact have stimulated ACP value-added exports to the EC, and so contributed to the economic development and diversification of the ACP states. In other words, does the Lomé Convention work? Is it effective in promoting the economic development and trade diversification of the ACP states? How will the formation of the Single European Market affect ACP prospects in the EC market?

EEC/ACP trade under Lomé

EC imports of all products from the ACP have fluctuated over the period 1970–87 (see Table 13.1). However, the general trend in the ACP share of all EC imports would seem to be slightly downwards. A similar pattern is evident for the ACP share of EC imports from all developing countries (Class 2). If Lomé had been effective, then the ACP share in total imports, and of all developing country imports, could have been expected to increase. The fact that it has not, indicates that the Lomé regime has not promoted ACP exports. What it has done is merely to protect the ACP market share without expanding it.

Similarly, the ACP share of total exports from the EC has been declining, as has the ACP share of EC exports to all developing countries (see Table 13.2). It could be argued that if the Lomé regime had worked, then the ACP share should have risen. This is because growth in ACP economies would stimulate more imports, and diversification in export products would require capital equipment which the EC could supply.

When EC imports of manufactured products alone are examined, the decline in ACP performance is striking (see Table 13.3). From an overall share of 5.49 per cent in 1970, the ACP share had fallen to 2.29 per cent in 1977 and 1.61 per cent in 1984 (the latest data available) and is probably even less today. The decline in ACP share of all developing country imports is just as startling. From 32.32 per cent in 1970 it fell to 12.32 per cent by 1977, and 8.30 per cent in 1984. At this general level, the Lomé regime has clearly not encouraged the growth and diversification of ACP value-added exports. As the decline in ACP share of all imports has been somewhat less than that of manufacturing products, the ACP reliance on traditional commodity exports has been even more reinforced.

Thus, poor performance by ACP exporters is also reflected in the growth rates for manufacturing imports, and all imports to the EC. While EC imports of all products from the ACP grew by 6.66 per cent over the period 1980–84, manufactured exports grew by only 2.62 per cent. The figures for Mediterranean countries were 14.69 per cent for all products, and 15.97 per cent for manufactured ones. At this general level it would appear that the Lomé regime has not worked as well as was hoped.

Table 13.1   *EEC (12) imports of all products*

	WORLD	CLASS 2	ACP	ACP % OF WORLD IMPORTS	ACP% OF CLASS 2 IMPORTS
1970	124,371	23,512	5,472	4.40	23.27
1975	263,229	60,745	9,715	3.69	15.99
1977	365,822	83,544	13,515	3.69	16.18
1978	387,096	78,756	12,719	3.29	16.15
1979	469,963	97,875	15,746	3.35	16.09
1980	557,746	129,233	20,744	3.72	16.05
1981	618,870	144,488	18,802	3.04	13.01
1982	672,187	144,765	20,140	3.00	13.91
1983	707,694	138,610	21,903	3.09	15.80
1984	809,357	151,769	27,749	3.43	18.28
1985	874,675	155,945	30,310	3.47	19.44
1986	796,005	107,663	19,575	2.46	18.18
1987	829,134	108,492	16,374	1.97	15.09

Source: Eurostat: External Trade Theme 6 Series A.

Table 13.2   *EEC (12) exports of all products (million ecu)*

	WORLD	CLASS 2	ACP	ACP % OF WORLD IMPORTS	ACP% OF CLASS 2 IMPORTS
1970	116,157	16,789	4,068	3.50	24.23
1975	249,184	45,679	8,772	3.52	19.20
1977	345,947	65,114	13,348	3.85	20.50
1978	374,530	70,278	13,623	3.64	19.38
1979	437,602	74,419	12,786	2.92	17.18
1980	497,137	89,296	17,048	3.43	19.09
1981	571,054	118,472	20,469	3.58	17.28
1982	626,652	124,416	20,222	3.23	16.25
1983	671,884	123,593	17,601	2.62	14.24
1984	776,772	131,136	18,069	2.33	13.78
1985	849,936	128,913	19,336	2.27	15.00
1986	806,958	107,602	16,049	1.99	14.92
1987	829,911	104,675	13,843	1.67	13.22

Source: Eurostat: External Trade Theme 6 Series A.

The impact of the Lomé trade regime

It was in the light of such poor performance on the part of the ACP that in 1984–5 the University of Kiel in West Germany undertook a comprehensive analysis of ACP–EC trade.[5] Its purpose was to analyse why the trade effects of the Lomé agreements seemed to be minimal for ACP states: in other words, why they met with little success with regard to export growth and

Table 13.3 *EEC (10) imports of manufactured products (million ecu)*
(SITC 5 + 6 + 7 + 8)

	WORLD	CLASS 2	ACP	ACP % OF WORLD IMPORTS	ACP% OF CLASS 2 IMPORTS
1970	25,588	4,495	1,452.8	5.68	32.32
1977	72,825	13,545	1,669.3	2.29	12.32
1978	82,218	15,118	1,698.0	2.07	11.23
1979	97,205	19,075	2,048.5	2.11	10.74
1980	115,353	22,783	2,672.9	2.32	11.73
1981	124,722	24,200	2,547.3	2.04	10.53
1982	136,721	26,557	2,481.1	1.81	9.34
1983	151,012	29,772	2,709.3	1.79	9.10
1984	183,039	35,729	2,963.8	1.62	8.30

Source: Eurostat (1986): Trade EC – Developing Countries & Manufactured products – Analysis 1970–1984.

diversification. The study also aimed to reveal which policy changes needed to be made in order to increase the trade effects of the preferences.

Its findings can be summarized as follows: •

(a)   The overall tariff margins of ACP states had been eroded. This was mainly due to the enlarged coverage of the GSP in processed agricultural products from 145 items in 1971 to 338 in 1983, and the depth of GSP-induced tariff cuts. However, preference margins may have been maintained by quotas and ceilings for non-ACP developing countries rather than by keeping the tariff margins constant.

(b)   It is the relative prices of ACP and non-ACP exports and their changes which influence ACP competitiveness in the EC. Attention should be paid therefore to those factors which influence the price level of exports.

(c)   External barriers to ACP growth were identified: for example drought, plant and animal disease, desertification and poor transportation infrastructure.

(d)   ACP policy-induced barriers to export growth and diversification were revealed: for example factor price distortions, high inflation, overvalued exchange rates and disincentives to the formation of indigenous entrepreneurship.

In 1988 the author led a study team in the United Kingdom which undertook further investigations into the extent to which ACP preference margins in the EC market had been eroded. They looked at whether the Lomé provisions have encouraged ACP exports of non-traditional products. The investigation was undertaken at the most specific product level for which data was available. The raw statistical data for the Reading Study were derived from Eurostat and consisted of figures for the value and volume of exports for selected products to the EC from ACP states, together with all developing countries as a group over the period 1976–86. The CET and GSP tariff rates and quotas for the years 1976, 1981, 1983 and 1986 had to be identified. The Lomé tariff rate applicable to ACP exports was zero throughout.

The purpose was to evaluate the impact which the provisions of the various Lomé Conventions have had on value-added exports from ACP states to the EC. This necessitated contrasting ACP exports since 1976 with an alternative situation, the so-called 'anti-monde'. The implementation of the Lomé regime had coincided with a period of extensive changes in the international trade scene. This made it relatively meaningless to use projections of pre-1976 trends as the alternative 'anti-monde' situation. Instead, ACP value-added exports to the EC were contrasted with those of the other developing countries (ODCs). While ACP states benefited from zero tariffs under the Lomé regime, the ODCs faced either higher GSP or CET rates, and/or quotas. If the Lomé Conventions had any impact on ACP value-added exports, it would be reflected in the level and share of these exports relative to those of the ODCs.

There were three stages to the analysis: firstly, the change in the ACP share of developing countries exports was compared with the change in the total level of exports for each product. The ACP share was said to have increased whenever the annual average rate of change for the ACP was greater than for the ODCs. Second, the change in the ACP share for each product was contrasted with changes in the level of tariffs or non-tariff barriers that the ODCs faced. Where these barriers fell, then it could be argued that ACP preferences had been eroded. Third, the change in the ACP share for each product was contrasted with its absolute level of exports compared with that of the ODCs. In this way, it is possible to reveal the extent to which ACP export performance had improved or worsened.

The methodology has three inherent problems.

(1)   For certain products, EC imports from the ACP and the ODC undoubtedly represent only a fraction of total EC consumption. This means that the ACP states and ODCs are competing not only with each other, but also with developed EC member states. This tends to confuse the overall picture.

(2)   For other products, the data base includes similar products which show contrasting export trends. This suggests that fluctuations in the exports could have arisen from 'product switching', rather than any distinct preference-related effect.

(3)   Finally, some of the trends may appear unusually favourable or unfavourable because of abnormally high or low figures in the base years 1976 and 1981 or because exports started only after 1976.

With due consideration made to these limitations, however, the analysis can indicate the impact which the Lomé regime has had to date.

Some fifty products were selected for analysis. They were the major value-added products exported by ACP states to the EC. As such it was anticipated that they would provide a representative indication of the impact the Lomé regime has had on ACP exports *vis-à-vis* those of other developing countries.

The first stage of the analysis revealed those products for which the ACP's share, by value, increased or decreased relative to that of the ODCs. For thirty-four products the total level of EC imports from developing countries increased, and so did the share for the ACP. For one product, unbleached cotton fabric 85–115 cm (CCT 550915), the ACP share increased even though the total level of imports of Class 2 countries declined. However, in the case of fourteen products the ACP share declined, although total imports increased. For the last product, preserved pineapples over 19 per cent sugar under 1Kg (CCT 200665), total imports fell and so did the ACP share.

The change in share can be broken down into two periods: 1976–81 and 1981–6. In the case of two of the products for which the ACP share fell for the whole period, this was because the annual average rate of change was especially low in 1976–81. In the period 1981–6, the exports increased faster for the ACP than for the ODC. The two products were cocoa paste not defatted (CCT 180310) and other goat and kid leather (CCT 410491). Conversely, for nine of the products, although the ACP share increased for the whole period under review, its rate of growth was lower in the second period than was that of the ODCs. Examples of this trend were cigars (CCT 240220), sisal binder twine (CCT 590431) and cotton gloves (CCT 600270).

The trends in the volume figures are the same as the value figures for nearly all the products. However, there were a few exceptions. For unbleached cotton fabric over 165cm width (CCT 550917), although in value terms the market size (total imports) increased as did the ACP share, in volume terms both fell. This may be a case of product switching, as the market size and ACP share increased in both value and volume terms for unbleached cotton fabric, 115–165 cm width (CCT 550916).

Another exception was other goat and kid leather (CCT 410491). In value terms total imports increased and the ACP share fell, but in volume terms total imports fell and the ACP share increased. This suggests that within the sector, the ODCs may be more specialized in higher-quality, higher-priced products, while ACP products are at the lower end of the market. This could be because ODC producers face a ceiling on the amount they can export. Therefore, they moved up-market in order to increase profitability.

Finally, for rum over two litres (CCT 220953), cigars (CCT 240220) and women's cotton blouses (CCT 61028X), although in value terms the total imports rose, in volume terms they fell. Nevertheless, the ACP share still increased. In the case of rum there may have been product switching as the value and volume of total imports of rum under two litres (CCT 220952) both increased.

Subsequently the change in the ACP share was compared with changes in GSP tariff rates and non-tariff barriers for the ODCs. Two basic questions were posed:

(a)   Did GSP barriers fall for those products where the ACP share fell? This would indicate preference erosion.

(b)   Conversely, did GSP barriers also increase for those products where the ACP share increased?

For one product, men's cotton trousers (CCT 61017X), the tariff change was not specified clearly enough to be included in this part of the analysis. Of the remaining fourteen products for which the ACP share by value fell, GSP barriers either remained the same or fell for ten of these products. In the case of clove oil (CCT 330123), the fall does not seem to have been due to preference problems. The principal exporter, Madagascar, shows some seasonal fluctuations in exports, suggesting that production problems are to blame for the fall in ACP share over the period. Also, as EC imports from ODCs tended to decline in the years in which ACP exports were high, responsibility cannot be attached to any particular preference problem.

In the case of sawn tropical wood over 1 mm (CCT 441455), the fall in the ACP share and the high rate of growth of ODC exports is partly explained by the very low ODC value and volume figure for 1976. Preference erosion is not to blame, particularly as barriers had been static for the whole period. For plywood, all the ACP exporters are in West Africa and there are considerable annual fluctuations, suggesting that production problems are the major cause of the fall in relative market share. However, ACP unit prices were some 80 percent higher than those for the ODCs. Although the barriers have remained constant, low price exporters from India and Indonesia in particular must pose a competitive threat to the ACP exporters.

Unwrought aluminium (CCT 760111) was one product for which the ACP performance has been particularly poor. Up to 1983 EC imports from the ACP states exceeded those from the ODCs, but they now have 55–60 per cent of the EC market. The sharp fluctuations in export volumes for Ghana and Surinam, two of three principal ACP exporters, suggest that there have been production problems. This was one product for which the ODCs have faced no barriers. With no market preferences from the Lomé Conventions, the ACP have been unable to compete.

For aluminium oxide (CCT 282011), although the ACP share has fallen, the ODC share is still negligible and there is no GSP rate. In the case of other goat or kid leather (CCT 410491), the ACP share only fell in value terms. GSP barriers have been constant, and, if it is the case that ODC producers have moved up-market, the ceiling has served to assist ACP exporters.

The final two products for which preferences were eroded were preserved pineapples over 19 per cent sugar under 1 kg (CCT 200665) and under 19 per cent sugar under 1 kg (CCT 200667). The ACP share fell as did that for the other two types of preserved pineapples. In the case of the latter, however, tariffs rose in 1981 and 1986 and went down in 1983.

Taking the four preserved pineapples products together, it would seem that preference erosion was probably not the only cause of the decline in the ACP share. There was some fluctuation in individual country exports, indicating that production problems may also have occurred. However, perhaps more so than for any of the other products, it is possible to argue that any benefits for the ACP from the Lomé regime have been eroded in the case of these particular products.

The four remaining products for which the ACP share fell all experienced an increase in GSP tariff rates. These were cocoa paste not defatted (CCT 180310), cocoa paste defatted (CCT 180330), pineapple juice with added sugar (CCT 200751) and pineapple juice without added sugar (CCT 200753). Clearly, for these products other factors were at work. The principal ones are probably production and/or marketing problems.

The results for those products where the ACP share declined must be evaluated against those for which the ACP share increased. For thirty-four of these products, the level of Lomé preferences actually fell, or at the very least, remained the same. Only one product, cocoa butter (CCT 180400), experienced an increase in preferences and an increase in the ACP share. This suggests that the Lomé tariff benefits have played a relatively minor role in determining the growth or fall in ACP exports of all value-added products. For many products, the ACP share increased while tariff benefits were being eroded. However, ceilings and quotas have remained on ODC exports, and it was principally these non-tariff barriers which may have protected the ACP market share.

The change in the ACP share for each product compared with its absolute level of exports was also examined. In so doing a major issue of underlying concern to the ACP states was addressed. They are the group of developing countries with a special relationship with the EC, yet do they command a greater share of EC imports than do the ODCs?

The results of the analysis revealed that ACP exports can be divided into four groups of products. The first group consisted of nine products for which ACP exports were higher than those of the ODC and the ACP share of the market was increasing over the period. Indeed for four of these products the value of ACP exports was actually lower than that for the ODCs in 1976. By 1986, however, ACP exports had overtaken those of the ODCs. The products were roasted nuts (CCT 200603), sawn tropical wood under 1 mm (CCT 441451), silk scarves (CCT 610610) and wooden kitchen furniture (CCT 940357). Yet, for eight out of the nine products GSP tariffs fell over the period. The exception was cocoa butter for which tariffs increased. The trend in the volume figures was the same except for silk scarves where the ACP share was lower. This suggests that ACP producers may export higher-quality scarves.

The second group consisted of seven products for which ACP exports were higher than those of the ODC, but the ACP share was declining. In the case of aluminium oxide (CCT 282011), however, the ODC share was negligible. Nevertheless, for the remaining six products, the ACP can argue that their earlier competitive position has been eroded. The poor performance of ACP exporters cannot, however, be blamed entirely on preference erosion. For three of the four products tariffs actually rose. It is true that for the other three, preserved pineapples (CCT 200665), clove oil (CCT 330123) and sawn tropical wood over 1 mm (CCT 441455), the tariff margin declined. Production problems and product switching in part could have also been responsible. Moreover, in the case of sawn tropical wood (CCT 441455) the ACP still have 83 per cent of the market.

Twenty-six products comprised the third group. In their case ACP exports were lower than those of the ODCs, but the ACP share of the market has been increasing. They could be judged as successes for the Lomé regime. The competitive strength of ACP exporters did not totally depend on tariff preferences as these have steadily been eroded. However, non-tariff barriers remain in existence.

The final group comprised eight products for which the ACP share was less than that of the ODCs, and was falling. Indeed, for preserved pineapples (CCT 200638 and CCT 200667) and unwrought aluminium (CCT 760111) the ACP performance has been so poor that its original position as the major exporter has been completely turned around.

Conclusions

The results of the study suggest that the ACP states can legitimately question the actual benefits of the Lomé regime when they still cannot meet the competition from other developing countries. The ACP are also justified in being sceptical of any future Lomé benefits as their preferences are being steadily eroded.

However, the fact that for thirty-five out of the fifty products, the ACP share has increased indicates that the Lomé regime has had some beneficial impact. What needs to be recognized is that the tariff and non-tariff provisions of the Lomé regime are not sufficient in themselves to increase and diversify ACP exports. If the domestic economies of the ACP states are not well managed, investment stimulated and production and marketing problems solved, then tariff benefits from the EC on their own will not secure long-term export growth and diversification.

After some twenty-five years of preferential treatment from the European Community, no ACP state has succeeded in fundamentally diversifying its economy and establishing a major dependence on value-added products. Mauritius perhaps represents the most successful ACP economy in that respect. At the same time, however, it also reflects the most profound ACP fear of latent EC protectionism with the possible imposition of so-called voluntary export restrains on Mauritian textile exports.

The evidence of this chapter suggests that the trade provisions of the Lomé Convention have had relatively limited impact on the ACP states. They do not appear to have generated the benefits which ACP states originally hoped for, particularly in terms of stimulating their value-added exports. Relatively limited trade diversification has occurred, and the evidence regarding erosion of the ACP preferential position in the EC market remains inconclusive. Other aspects of the Lomé Convention are doubtless of greater significance in terms of the longer-term economic development and diversification of the ACP economies. However, the advance of the Uruguay Round of trade negotiations, especially the EC's offer on tropical products, means that ACP states urgently need ameliorating measures to protect their position.

The next Lomé Convention (Lomé IV) is likely to operate in a world increasingly dominated by trade in manufactures. ACP exports of value-added products are likely to find themselves under pressure from competitors in other parts of the developing world. ACP economies, heavily dependent on exporting raw materials, will need to be bolstered by practical and effective schemes to promote structural adjustment and viable economic development.

In the context of the Lomé IV negotiations and moves towards 1992 the ACP states have already tabled a number of policy objectives:

(1) The preservation of their preferences in the EC market;
(2) Clear, quantifiable compensation measures from the EC to offset the erosion of ACP preferences;
(3) A commitment from the European states recognizing that the ACP states have lost ground in the EC market due to GSP concessions;
(4) Increased resources to be invested in marketing and product diversification in the ACP states;
(5) The ACP states need to harmonize their own negotiating position within the Uruguay Round and promote much stronger collective action.

The ACP states are looking for a specific Lomé approach to their structural adjustment as distinct from the programme operated by the IMF and the World Bank. These structural adjustment programmes, allied with balance-of-payment support measures, will need to be integrated with a more energetic focus on product development and marketing assistance for value-added products from the ACP states. To date, there is relatively little evidence to prove that the

Lomé Convention has worked in terms of its goal to develop and diversify the ACP economies.

References

1. Michael Prest, 'A Trade Partnership of Equals', *The Independent*, 16 February 1989.
2. Christopher Patten, 'The Lome Renegotiation', *EEC London Press Service*, 9 February 1989.
3. Op. cit.
4. Carol Cosgrove Twitchett, *A Framework for Developement: the EEC and the ACP*, London, 1981.
5. Jamuna Agarwal, Martin Dippl, and Rolf Langhammer, *EEC Trade Policies Towards Associated Developing Countries: Barriers to Success*, Tübingen, 1985.

14 Development assistance under Lomé IV: politics or economics?
Carol Armistead Grigsby

Introduction

Many of the Contributors in this volume are European scholars. I, on the other hand, approach the subject of the European Community's development policy as one familiar with the issues facing the aid community and curious about how the EC is going to face up to them. My chapter will be structured as follows:

First I shall briefly set the stage, revealing my biases and examining the European Community's place in the spectrum of aid donors. I shall go on to describe the problems all aid donors are grappling with in Africa, where the EC directs most of its aid under the Lomé Conventions. Next, I will look at the constraints on, and the potential for, the EC to become a meaningful actor in the provision of development aid. I will conclude with a few thoughts about the directions the EC might take under Lomé IV.

Before continuing, I would like to make two things clear. First, I intend to focus on development aid under Lomé, not on the trade or other aspects of the agreement. Second, when I refer to the EC, I am referring to the institution founded with the Treaty of Rome and headquartered in Brussels. There is a clear distinction to be drawn between the EC, whose development aid is administered by the Eighth Directorate General of the Commission in Brussels, and the individual aid programmes of the twelve European governments belonging to the EC. Indeed, my main message is the following: Europe as a whole, including the EC and its members, is the largest source of development aid going to Africa. However, these different European aid programmes have not been coordinated and therefore do not provide a meaningful counterweight to the influence of the World Bank on that continent. The EC Commission, now trying to break free of its reputation as a purely political entity, has its work clearly cut out for it, due to both external constraints and to the preferences of the individual European members which fund it and determine its policy directions.

Negotiations have recently begun for the fourth Lomé Convention, Lomé IV, which will take effect in 1990 and run through 1995. Although the Lomé renegotiation is not formally linked to the landmark 1992 date, the coincidence in timing heightens the visibility of Lomé and guarantees a 'fresh look' at its provisions. With a mandate from its twelve members to proceed, the EC

Commission is now pursuing discussions with the sixty-six African, Caribbean and Pacific countries—the ACP. Forty-five of the ACP countries are in Africa, and include all of Sub-Saharan Africa except Namibia and South Africa.

The EC alone, without including its members' bilateral programmes, has consistently been one of the largest donors in Africa.

— In 1987, according to the OECD, the EC was second only to France in disbursements to African countries. The EC's disbursements exceeded even those from IDA, the World Bank's concessional lending arm. Although the level of assistance has not been so high every year, the EC is consistently one of the top five donors on the continent.

— For the moment, others are also active in Africa. The World Bank is dedicating large amounts of aid to Africa due to its debt and development crisis; Japan, under pressure to find outlets for its substantial surpluses, is increasing its assistance; and the United States is responding to a strong humanitarian outcry from Congress and the public to aid the continent.

However, Europe has historic ties which ensure that it will remain engaged in Africa's development long after other countries have pulled back. Counting aid from EC member countries, Europe as a whole accounts for about half the resources going to Africa. If the general trend toward stronger collaboration among all donors, and the beacon of 1992, can catalyze closer coordination of EC and European aid programmes, Europe has the clear potential for an even greater role in African development.

The present context of development aid

Now, I would like to take a look at the context of development aid in which Lomé IV negotiations are taking place. The prominent factor before the aid community at this time is the problem of developing country debt. While Latin America is the focus of most of this attention, the ACP countries are far from immune. Nigeria and the Ivory Coast, two of the seventeen major debtor countries, belong to the ACP. But even for low-income countries whose debt does not present a threat to the international financial community, the burden of debt at the national level is severe. Between 1985 and 1988, twenty of thirty-four low-income African countries rescheduled their official debt owed to developed country governments, and yet debt service ratios still average around 35 per cent.

In part due to these debt burdens, many developing countries have begun to question the policies they have pursued since independence. By the mid-eighties, many African countries were recognizing that basic tenets of their economic policies were ill-founded. At the UN Special Session on African recovery in 1986, African countries made an historic agreement with the industrial countries, agreeing to reform their economies along more market-orientated lines in exchange for needed financial support.

This need to re-orient economic policies has led to widespread use of adjustment assistance in Africa. Adjustment aid is not linked to specific investments, but provides budget support in exchange for promises by the government to make its economic policies more efficient. Structural adjustment operations support macroeconomic policy changes, while sector adjustment operations are limited to reforms in a single sector.

Donors, too, have taken responsibility for past behaviour by taking action to coordinate their programmes of assistance in developing countries. This was overdue, since over-eager donors pursuing commercial advantage or the latest development fad account for many of the 'white elephant' projects dotting the African countryside. Coordination becomes even more important, though, when what is at stake is the economic programme being followed by the government of a developing country.

Leadership in adjustment lending and its coordination has been provided by the World Bank. From an average of 15 per cent in the early eighties, adjustment loans now account for 35 per cent of Bank commitments in Africa. The Bank's Special Fund for Africa and its successor, the Special Programme for Africa, will have attracted about $7 billion in commitments from other donors to assisting African economic programmes by next year. The IMF's Structural Adjustment Facility, also focused on Africa and closely programmed with the World Bank, has recently been expanded with $6 billion in new donor resources. The OECD estimates that about half of aid to Sub-Saharan Africa from all sources is now related to structural reform. Ghana, which has undergone an economic transformation over the past four years, is a striking example of the donor response to new economic attitudes in Africa.

The EC: facing the challenge

So just where has the EC been while all this activity has been going on? To be quite honest, a popularly held view in the aid community is that the EC is at best innocuous, and at worst working at cross purposes with other donors in Africa.

In fact, though, the EC has not stood still. Although aid under Lomé traditionally has focused on projects, the EC has found several ways to support international efforts to assist economic adjustment in Africa:

(1) First, under the third Lomé Convention, covering the period 1985–90, two significant measures were taken. The EC greatly strengthened its Indicative Programmes of Aid, under which it negotiates with the

recipient government the agreed sectors to receive financial assistance, the directions of the recipient's economic planning and an approximate timetable for instituting needed reforms. Also under Lomé III, the EC has undertaken sector-wide operations on a much larger scale.

(2)   Since Lomé III began, other measures have been instituted. In 1987, the EC programmed over 500 million dollars to accompany the World Bank's Special Programme for Africa. The EC also recently began programming its food aid in close conjunction with Lomé development aid. EC food aid, which does not come out of the Lomé budget, had previously been managed in a separate directorate of the Commission from that administering aid under Lomé.

The EC Commission has more ambitious plans for the period of Lomé IV. It has asked the twelve to agree that the EC should undertake structural adjustment operations, and they have authorized it to try to negotiate this as part of the Lomé IV agreement. So on the face of it, the EC appears to be moving toward a consensus with the international aid community. Caveats are in order, though. The Commission has posed basic questions about the World Bank's advice to adjusting countries. Documents prepared by the Commission point to a number of areas in which it believes a different approach from that followed by the World Bank is needed. The Commission's concerns include the following:

(1)   First, the Commission thinks that the so-called 'adjustment fatigue' taking hold in some African countries springs from a failure by donors to accommodate 'home-grown' solutions to economic problems and consider alternative approaches. Therefore, it contends that adjustment reforms should take greater account of a country's political and social constraints, not imposing any particular development model on recipient governments. This implies slower removal of subsidies, greater tolerance for parastatal enterprises, and the like.

(2)   Second, the Commission argues that the recipient government must be closely involved from the start in designing the adjustment programme. According to this view, the World Bank tends to design the adjustment programme and merely present it to the government as a *fait accompli*. In principle at least, the EC has an advantage over the World Bank here. The European Community has always had extensive local representation in Africa, and now has offices in thirty-five African countries. By comparison, the World Bank is a johnny-come-lately to field representation. The long-standing presence of in-country staff familiar with local conditions can be an advantage in designing realistic development programmes. In addition, the Commission's delegations in developing countries enjoy somewhat greater decision making authority as a general rule than do World Bank field offices.

(3)   Third, the EC Commission supports food security strategies that emphasize food self-sufficiency in African countries. This runs counter to the World Bank approach to food security which includes less scope for subsidies, import restrictions and other measures to protect national productive capacity.

(4)   Fourth, the Commission argues that adjustment packages, which usually entail budget cuts, should ensure that certain basic subsidies or social services will not be the first expenses cut from the government budget. Thus, measures to mitigate unfavourable social effects of the adjustment process need to be built into adjustment programs from the start.

(5)   The fifth prominent EC concern about the Bank's approach to adjustment lending is that adjustment programmes must have a longer time horizon, with greater certainty as to funding amounts.

The EC is not the only one to raise such concerns. In fact, to the extent that some rethinking of adjustment lending is already taking place, some of these concerns may be assimilated into present adjustment efforts without much difficulty. The recent World Bank review of adjustment experience highlighted the need for strong government commitment to the adjustment programme and for incorporating social measures into adjustment programmes. However, other elements in the EC Commission's approach, such as greater tolerance for alternative economic models and support for food self-sufficiency measures, will not be so easily harmonized.

If one takes at face value the mandate the EC membership has given the Commission to negotiate in this area, all would still seem to be well. However, in reality the EC faces both internal and external challenges on the road to adjustment. To some extent, these challenges arise from the internal dynamics of the Lomé system itself. First is the historical context of Lomé. Lomé I built on the two Yaounde Coventions, which ensured preferential treatment for France's former colonies in the early years of the EC. The first Lomé Convention was established after the United Kingdom entered the EC in 1973. At that point, the key objective was to incorporate Britain's ex-colonies into the existing special arrangement for Francophone countries. Structurally, Lomé reflects a political approach to relations with developing countries, which presents further obstacles:

— The Conventions take the form of a treaty between the EC member countries on the one hand and the ACP countries on the other. In theory at least, the two groups are equal partners in the agreement. Many European scholars have warned against taking this too seriously, and the ACP countries criticize the erosion of their bargaining power with each successive Lomé Convention. Still, there remains at least a vision of partnership at the core of Lomé, and this implies a different power relationship between donors and recipients than that found in the World Bank and similar institutions where decision making power is firmly concentrated in donor hands.

— The logical manifestation of this approach is the way in which EC resources are allocated among recipients. At the outset of each agreement the exact amount due each ACP country is set, so that it can be sure what it will receive over the period. These entitlements preclude any subsequent decisions to cut off or reduce funding for a particular country. Shifts in government priorities, or changes in governments themselves, do not jeopardize these guaranteed resources. To be fair, the Commission can and does delay disbursements in some cases, but the inability of the EC to withdraw its support even in grievous cases removes any leverage it might have in negotiating with developing country governments.

So far the EC's donor governments and its ACP recipients, the political format of the Lomé arrangements constrains the Commission in its efforts to be taken seriously. But even if the EC can overcome this, it faces other challenges.

First and foremost the EC encounters a paradox in its timing. Just as it is preparing to move into this area, the tide may be starting to turn against adjustment lending. Even some of the major European countries are beginning to question both the effectiveness of past adjustment lending and the need to continue present levels of such lending. Adjustment lending originally was seen as a fairly short-term effort to help get countries' economies back on track so that productive investments could resume. One can already detect 'aid fatigue' setting in as donor countries envisage another decade of balance-of-payments support in exchange for policy changes which might, or might not, prove effective or sustainable.

Second, and also crucial, to be a significant player in adjustment lending takes money, and it is not clear the EC members will provide it for this purpose. In Lomé III, the size of the aid package was barely maintained in real terms. A further complication this time is the fact that the ninth IDA replenishment in the World Bank is being negotiated at the same time as Lomé IV. This places Lomé funding and IDA funding into direct competition for European development resources. It is true that the Commission has its mandate to try to negotiate a programme of structural adjustment under Lomé IV. However, in the final analysis, who knows whether EC members will trust the Commission to protect their investment by enforcing adherence to policy reforms? IDA, as part of the World Bank Group, still is seen even by the Europeans as a more effective enforcement mechanism. This goes back to the fact that most European governments persist in viewing Lomé as a political link to former colonies not as a serious development institution.

EC: the potential

The EC has the potential, and the resources, to play a strong role in African development. But it has to resolve—first with its own membership, then with other donors—the differing perspectives on its appropriate role. This leaves the EC with two contrasting options.

(1) If the Commission had its way, it would probably like to present an alternative for countries uncomfortable with World Bank conditions. Such an approach would probably accommodate ACP views. 'Alternative adjustment' under this scenario would consist of a gentler pace of reform, with, for example, more tolerance for subsidies, slower price decontrol and greater tolerance for parastatal enterprises. It could also mean explicit programmes of food self-sufficiency which could quickly bring the EC into conflict with the World Bank.

Simply by providing an alternative, this approach would undermine Bank-led efforts and bring the two institutions 'eyeball to eyeball' far more often. Partly for this reason, this route would require far greater support than is now the case from European countries—in the form of resources, moral support and a willingness to programme their own bilateral aid in connection with the EC rather than with the World Bank. Only if the EC had the full backing of its members could it serve as a meaningful counterweight to the Bank. This is a long shot precisely because of the powerful pull which the Bank exerts over the whole development field.

(2) The other option for Lomé would be to incorporate adjustment lending only to the extent that it can be fully integrated with current World Bank efforts. Like the resources now programmed in conjunction with the Bank's Special Programme for Africa, the EC would leave the design and conditions of the adjustment operations to the Bank. The EC would be providing resources fundamentally to enhance Bank programmes. Possibly, the Commission could negotiate with the Bank to obtain slight variations on its present adjustment model in specific country cases, but by and large it would accept Bank judgements.

This would give legitimacy to the EC's programmes by maximizing EC participation in international coordination efforts and would also help dispel impressions of the EC as a purely political institution. However, it would mean that the EC's rigid system of resource allocation would have to become more flexible so it could follow Bank programmes where they led.

In my judgment, the EC will ultimately opt for the latter approach. It will do so, first, because it is the most likely route to obtain European countries' agreement to go into this area. Second, it will not want to be accused of swimming against the tide when coordination is the order of the day. Third, it will do so because it knows that the Commission does not yet have staff with the right qualifications to develop economic advice that can inspire the confidence of African governments and other donors. However, the EC might try to exact an exchange from the Bank by influencing somewhat the design of its programmes in particular countries. In selected cases, this might even include trying a programme or two of its own in cases where the Bank may fear to tread. The EC's success in these cases will determine whether it later has the courage to branch out on its own. For now, though, the cautious route will dictate sticking close to the herd.

At the beginning, I pointed to the potential which Europe has for taking the initiative in Africa. Now I have concluded with a fairly downbeat assessment of the chances for EC leadership on development in the region. The EC may have something distinct to offer, though, as it ventures into its new role. I have mentioned the political foundation for the partnership concept under Lomé. If this could be given a more practical meaning—for example, through a more collegial approach to designing adjustment programmes *with* the African governments which have to implement them—then the EC could bring something worthwhile to the process. Possibly, the EC could turn partnership into the hallmark of its participation in these programmes.

Over the longer term, the EC does stand a chance of becoming that leader among donors in African development. First, though, it has to bide its time, build up its competency and credibility and, above all, gain the respect of its own membership. Only then will it become a force to be reckoned with.

Note

1. *Carol Armistead Grigsby* works for the U.S. Agency for International Development in Washington, D.C. Her views are entirely her own and do not represent U.S. government policy.

Index